STINGERS

Vietnam War - Helicopter Gunships

Fred Allen
Crew Chief / Door Gunner

ABOUT THE COVER

September 24, 1970, Quang Ngai Province, South Vietnam. It was near dusk with light rain. We were somber, returning from an unsuccessful rescue mission.

Our Cobra wing ship had proceeded back to base ahead of us. Old 523 was low on fuel. As I surveyed the instrument panel between the pilots I noticed the torque meter registering forty pounds. That was the highest load level I had ever seen.

I keyed the intercom for the Aircraft Commander, "Hey Smitty, the old bird is pulling forty pounds." The pilots looked at each other and smiled.

The next moment, everything exploded. Heavy machine gun enemy tracers burst through the pilots' radio console and tracked through the door gunner cabin, piercing the fuel cell. Another enemy bullet hit the bottom of the ship and broke through the deck about six inches from my right foot. Shrapnel sliced into my leg and hand. The Aircraft Commander started firing rockets. The door gunner and I returned fire.

As I leaned out of the ship, I was hit by flames.........

CONTENTS

Title Page
About the Cover
Introduction
Copyright
Helicopter Air Crew Monument 1
Dedication 2
Note from the Author 3
Chapter 1 Tough Start 4
Chapter 2 Welcome to Military Hell 8
Chapter 3 America's Vietnam War 10
Chapter 4 Preparing for War 12
Chapter 5 Hunter Army Airfield 18
Chapter 6 Officer Candidate School 23
Chapter 7 Good Morning Vietnam 26
Chapter 8 The 116th Assault Helicopter Company 32
Chapter 9 My First Mission 46
Chapter 10 Stinger Memorial Service 53
Chapter 11 First Kill 55
Chapter 12 Lovers and Water Buffalo 61
Chapter 13 Cu Chi Incoming 64
Chapter 14 Aircraft 523 Gets Hit 67
Chapter 15 Combat Mentality 72
Chapter 16 Scramble to Support 75
Chapter 17 Reassignment to I Corps 80
Chapter 18 Kham Duc 89

Chapter 19	The Stingers and My Lai	92
Chapter 20	174th AHC Sharks Initiation	96
Chapter 21	Tam Ky Insertion	100
Chapter 22	Crash and Burn - Aircraft 523	105
Chapter 23	Massive Enemy Force	114
Chapter 24	Sergeant Curtis McKinley	117
Chapter 25	Addiction and Breakdown	120
Chapter 26	Spray Mission Agent Orange	124
Chapter 27	SP4 Tony Bremer - Silver Star	128
Chapter 28	Dear John	131
Chapter 29	Night Mission - Hawk Hill	133
Chapter 30	Yellow Jacket Down	138
Chapter 31	ARVN LZ Insertion	142
Chapter 32	Donut Dolly Run	146
Chapter 33	CW2 Lou Baldassaro	148
Chapter 34	Jesus Nut	150
Chapter 35	Fire Support Base Maryann	154
Chapter 36	First Female Enemy Kill	157
Chapter 37	LRRP Team Extraction	161
Chapter 38	Vengeance	164
Chapter 39	Turkey Shoot	168
Chapter 40	Change of Heart	172
Chapter 41	Refusal to Kill	174
Chapter 42	Catharsis	177
Chapter 43	Reassignment	179
Chapter 44	Sergeant of the Guard	181
Chapter 45	Friendly Fragging	184
Chapter 46	No Place Was Safe	187

Chapter 47 Charge of Quarters 188
Chapter 48 Second Fragging 190
Chapter 49 Lam Son 719 192
Chapter 50 The ARVN Retreat 204
Chapter 51 The 116th Returns to Chu Lai 208
Chapter 52 Freedom Bird 210
Chapter 53 Soldier's Aftermath 215
Epilogue 219
Thank You! 221

INTRODUCTION

The author presents this book based on his personal experiences in the Vietnam War. His intent is to portray his perspective as an 18-year-old Stinger Gunship Crew Chief/Door Gunner.

He wants the reader to experience what it is like to kill dozens of enemy combatants and collateral, non-combatants. How to live with the frequent near-death experiences and the constant high probability of being killed.

Endure the frequent loss of fellow soldiers in combat. Witness a young soldier losing all sense of humanity as he transforms into a warrior that thrives on killing.

The real-life incidents portray a platoon of Huey UH-1C gunships, called the "Stingers" and its elite, motivated crews.

As a component of U.S. Army assault helicopter companies, the "guns" protected the Huey "slicks" as they inserted or extracted infantry troops from landing and extraction zones.

The gunships regularly conducted "search and kill" missions.

They were often scrambled to provide lifesaving protection for infantry units threatened by enemy forces.

Gunship crews were among the most lethal pilots, crew chiefs, and door gunners in Army aviation. Some gunship crew chiefs had over 400 hundred personal kills.

An intense warrior mentality was crucial to mission success and survival. Killing was a way of life in the guns.

One of the most decorated and experienced aviation combat units in the Vietnam War was the Stingers' parent, the 116th Assault Helicopter Company, known as the "Hornets".

Follow the operational strategy as Army Command reassigns

the Hornets from III Corps, near Saigon to Chu Lai in I Corps under the command of the 23rd Infantry Division, known as "Americal".

Experience the rapid increase of lethal enemy encounters.

Americal imposed severe constraints on the Stingers in the aftermath of the My Lai massacre. The Stingers and slick platoons of the 116th fight on.

The young combat veteran becomes "short" with just a few months left in his tour of duty. Americal reassigns the 116th to Quang Tri, one of the most dangerous places in Vietnam.

The mission was to support the Laotian operation, Lam Son 719.

Quang Tri was less than fifteen miles from North Vietnam. The enemy rocket attacks on the Quang Tri base camp were relentless. Soldiers became shell-shocked.

War memories would haunt them for the rest of their lives.

The day finally came. The young warrior boards the Freedom Bird for the flight home. As the eastbound airliner crossed the border of South Vietnam, the returning soldiers break out into celebratory yells of relief.

The onboard jubilation is short-lived. War damaged soldiers experience sudden decompression from the constant lethal threat endured for months on end. Soldiers begin to emotionally break down. The civilian flight attendants rush to calm the suffering heroes.

War news headlines are articulated in real-time to frame the author's story.

The author attempts to portray a balance of bravado militarism with the constant dilemma combat soldiers faced in Vietnam.

It is a heroic endeavor to believe in upholding nationalistic pride with high principles of duty, honor, and country. But when combat soldiers lose America's support; when military morale and leadership falters; when the cause for war becomes questionable; a soldier has to find his own way to persevere and survive.

Paperback Soft Cover Edition - ISBN: 9780578317809

eBook Kindle Edition - ASIN: B09C3V2PWG

ebook Global Edition - ISBN: 9798201476588

Author Contact: USABooks1@aol.com

Cover design by: Tatiana Vila - Vila Design

Printed in the United States of America

HELICOPTER AIR CREW MONUMENT

Arlington National Cemetery

Nearly 5,000 Pilots, Crew Chiefs and Door Gunners Killed in Action

Created through the efforts of Vietnam Helicopter Pilots Association

DEDICATION

In honored memory of the 58,220 U.S. soldiers, airmen, and sailors who made the ultimate sacrifice and over 2,600,000 that lived with the war's mental and physical wounds.

Right or wrong, our Nation was at war.

Young Americans stepped up to serve their Country.

Vietnam Veterans Memorial

Washington D.C.

NOTE FROM THE AUTHOR

Actual events inspired the stories I describe. I portray incidents in a manner that is consistent with the reality of that time and place. I have changed the names of fellow crewmen and soldiers to protect their privacy or memory.

Writing this book has been a lifelong endeavor. I kept notes, as memories would surface over the years. As I approach seventy years old, I find I can now reflect on my war experience with greater clarity and perspective. The memory of constant lethality is clear.

The post-traumatic stress has gradually subsided, enabling my telling of this story. Perhaps this book is my catharsis.

Many of the photographs were taken with an instant Polaroid camera in 1970 -1971. The half-century-old images were digitized from the original prints.

The passage of time is claiming the Vietnam generation of America's remaining combat soldiers. May this book help to provide future generations with a unique insight into the Vietnam War. Perhaps history will preserve the memory of the brave young men in helicopter gunships who faced frequent lethal engagements in the service of their Country.

This is a graphic account of helicopter gunship operations and life in an active war zone. Some readers may find the content disturbing.

You can take the warrior out of the war, but you will never take the war out of the warrior.

Specialist Fifth Class Fred Allen
U.S. Army Aviation

CHAPTER 1 TOUGH START

It was cold and overcast on January 1, 1966, in Wichita, Kansas. My brother and I were waiting for our Dad to arrive from Cushing, Oklahoma, to pick us up after a holiday visit with our Mom and two sisters. Two years prior, Mom and Dad had divorced and settled in with new companions. They sent my younger sisters to live with Mom. My younger brother and I remained with Dad.

About mid-afternoon on that New Year's Day, we became worried about Dad not showing up. He was not answering his phone. In those days, there was no voice mail, answering devices, or mobile phones.

It was a three-hour drive from Cushing. I was ready to leave. As a fourteen-year-old, I did not get along well with my step-dad.

Then the call came. Mom answered the phone. It was Uncle Bob, our Dad's brother. It surprised Mom, since Dad's side of the family had no contact with her after the divorce. Bob asked, "Have you heard about Roy (our Dad)".

Mom hesitated before responding, "Well, no. We have been expecting him to pick the boys up."

Bob delivered the news that changed our lives forever.

Just after midnight amid New Year's celebrations, Dad and his female companion were driving on a two-lane highway west of Cushing. As you would drive up a hill, oncoming traffic would spring up at the top of the hill.

A drunk teenage driver crossed the centerline and appeared in Dad's lane atop the hill. The collision was head-on. Dad's companion suffered injuries. The drunk teenager was unscathed.

Seat belts were not being used. The steering wheel crushed

Dad's chest as the collapsing doorpost caved in his head. He died soon after he arrived at Cushing Memorial Hospital.

It became a different world. The grief was debilitating. The uncertainty of what would happen to my brother and me was frightening. Our family was low income, so options were limited.

After Dad's funeral, I realized I had to become a man. I had always been a self-starter, with various entrepreneurial ventures from an early age. I had my Dad's likeability. I was daring, which did not always serve me well. At age fourteen, it was time to find out if my attributes were going to make me or destroy me. I found new strength from tragedy.

They decided I could choose where I wanted to live. At first, I wanted to stay with my paternal grandparents in Cushing. Instead, I elected to move to Wichita with my brother, joining our two sisters, Mom, and step-dad.

We received a small insurance settlement from the drunk teenager's family for myself and my three siblings.

At first, life was a little rocky. I guess I emitted a sense of vulnerability after losing my father, which did not play well at my new school. Kids were quick to pick up on that vulnerability and made life pretty miserable for me.

Those were formative years. I had no choice but to get tough and fend for myself.

As the next few years wore on, the stress of living in a cramped mobile home with four kids, two adults, and tight finances increased. I became more rebellious and started down a dark path with the wrong people. There were a couple of brushes with the law. At age sixteen, I quit High School.

A few months later, Mom gave me an ultimatum; turn myself around or she would ask the court to send me to a reform school.

Mom's "tough love" was difficult for me to understand. She provided the impetus to motivate me and start building a positive life.

I found an opportunity in a classified ad. A contract operator needed truck and combine drivers to harvest wheat, oats, barley, and flax crops for farmers and landowners. The job started during May in Texas and would end around September somewhere in North Dakota, or possibly Canada.

It sounded like a grand adventure. I interviewed for the job. After a few days, I got the call. They hired me.

It turned out to be a large-scale operation. There were six brand new, very expensive harvesting combines, six grain trucks, and two support vehicles. I was the youngest of the fourteen-man crew.

My first assignment was driving the grain trucks. I would pull up to a rolling harvesting combine in the field and position the empty truck bed under the protruding grain auger. As we rolled up and down the field rows, in perfect tandem, the combine auger would gradually fill my truck bed with grain. I would pull a tarp over the payload, so the grain would not escape in the wind. I would race to the grain elevator in a nearby town and offload the truck.

That cycle repeated several times a day. After about a month, they promoted me to combine operator. I discovered what hard work and long days were all about. It sped up my maturing process. Working on a team to achieve time-sensitive objectives helped me to develop skills that would give me a head start in future endeavors.

Working in the fields of America's heartland fostered a deep appreciation of our country and hard-won freedom. The farmers and most people you would meet in the great American plains states were decent, hardworking patriots.

I remember the end of a long summer's day, watching six large combines in staggered formation in the distance, rolling along in a massive wheat field, silhouetted by the setting sun. The scenic beauty and industrial power inspired me with pride to be an American.

We completed our last job in North Dakota, in early September 1968. After a month back in Wichita, the custom harvesting operator hired me back to cut soybeans in north Louisiana. We finished up those contracts in early December.

After the holidays, I pondered what my next move would be to further my life. My experiences traveling the Great Plains, and then being exposed to the uniqueness of Louisiana, had deepened my love for America.

As a fatherless young man, I considered myself as America's son. The concepts of duty, honor, and country inspired me.

I became intrigued about joining the military. I always loved flying. I had flown in small airplanes with a couple of people. I wanted to become part of aviation. The military seemed to be a good option.

I was aware of public dissent during the Vietnam War. I did not understand what that was all about. Most of us characterized the protesters as a bunch of drugged-out hippies attacking our country and disparaging patriotism.

At age seventeen, with Mom's written consent, I joined the United States Army and requested aviation. My test scores were high and qualified me for any enlisted military occupational specialty.

I shipped off to Fort Leonard Wood, Missouri, for Basic Training.

It was February 1969.

CHAPTER 2 WELCOME TO MILITARY HELL

Army Basic Training is no picnic. Winter in Missouri can be arctic cold. Twenty-mile marches, in frigid weather, caused water canteens to freeze. Bivouacs in subarctic temperatures will toughen up the weakest person.

Transitioning from a civilian to a soldier is complex. You go through psychological changes without realizing what is happening. Foremost, you become acclimated to accepting and obeying authority with severe repercussions if you refuse.

They indoctrinate recruits in the strict Uniform Code of Military Justice. Drill Sergeants tear down independent personas with constant verbal thrashing coupled with intense physical activity. The recruits undergo strenuous daily field and classroom training, learning to be a soldier.

Killing people is not an ability natural to most civilians. How does the Army take a teenager and transform him into a killer?

Sensitive to reports that only a small percentage of soldiers in World War II and the Korean War could kill, the U.S. Army developed effective methods of preparing young impressionable civilians to kill people.

Recruits endured psychological rhetoric to kill, coupled with acquiring killing skills. The Army doctrine reinforced the concept of patriotic pride and duty.

The regimen appeared to have worked. Some studies reported that 85% of American soldiers could kill in the Vietnam War.

We enhanced bayonet training with impassioned cries, "KILL-KILL-KILL" as the recruit plunged the bayonet into the chest of an "enemy" dummy.

Cadence songs during marches that focused on killing Viet Cong or "Charlie", reinforced the drive to accumulate enemy body count.

On the rifle range, a Drill Instructor, using a watermelon, demonstrated how a human head explodes when hit with a military-grade bullet.

As recruits became soldiers, the drive to kill Viet Cong became ingrained. The ability to withstand enemy fire, incoming rockets, daily blood, and gore could not be taught. It had to be experienced.

I proved to be an excellent marksman with the M-14, M-16 rifles, and the M-60 machine gun.

One of my extra duties was to work in the unit's main administrative office, commonly referred to as "orderly room". I seemed to have a knack for organization and detail. That experience proved to be quite handy in the following years.

Our training company had over 200 men. Not all recruits made it through training. Commanders discharged some out of the service. Two men contracted spinal meningitis and died.

Upon graduation, we received orders for our Advance Individual Training (AIT). My dream became my reality. I received orders to report to Helicopter Maintenance / Door Gunner School in Fort Rucker, Alabama.

I believe the Army gave me purpose and direction. I learned the importance of mission accomplishment under extreme conditions.

CHAPTER 3 AMERICA'S VIETNAM WAR

November 21, 1967- *General William Westmoreland, Commanding General U.S. Forces—Vietnam, addressed the National Press Club. He made a statement regarding the outlook of the Vietnam War, "We have reached an important point where the end comes into view".*

January 31, 1968 - *The enemy's major Tet offensive changed the dynamic of American public opinion. Militarily, the enemy's countrywide attack was soundly defeated. Unfortunately, the earlier positive outlook expressed by General Westmoreland just two months before lost credence.*

February 27, 1968 - *Walter Cronkite of CBS News, considered being the most trusted man in America, stunned the Nation with his report, now immortalized as "We are Mired in Stalemate", basically suggesting the Vietnam War was unwinnable. President Johnson, after hearing Cronkite's report, is said to respond, "If I have lost Cronkite, I've lost the country". Indeed, the anti-war protests increased in ferocity.*

March 16, 1968—*The My Lai "massacre" occurred. An infantry unit belonging to the Americal Division killed over 500 unarmed men, women, and children in a village in Quang Ngai Province. They believed the area to be a Viet Cong stronghold. When the news broke in 1969, it caused international outrage. The American public's opposition to the war increased. The negativity contributed to declining military morale.*

In August 1970 - The 116th Assault Helicopter Company and its Stingers gunship platoon were assigned to Chu Lai, just north of My Lai.

November 3, 1969 - *President Nixon laid out a plan for the withdrawal of American troops. America's official involvement in the Vietnam War was in its fourth year. The American people's tolerance wore thin despite declining casualties. Confidence in government was eroding. The people's anger turned against everything about the war, including the warriors sent to fight.*

Fatal U.S. war casualties hit 11,780 in 1969, 6,173 in 1970, and 2,414 in 1971, according to the Department of Defense Casualty Analysis System. Annual fatal casualties were falling commensurate with declining troop strength. The percentage of fatalities did not change. The average annual death toll percentage of U.S. troop strength for the three years was 2.1%.

Of the 58,220 U.S. servicemen killed in Vietnam, nearly 5,000 were helicopter pilots, crew chiefs, and door gunners.

Over 100,000 helicopter crewmen served in Vietnam. The death toll percentage was near 5%, more than twice as fatal as the death rate for all soldiers, airmen, and sailors.

Gunship crews suffered an even higher, fatal casualty rate.

Military planners had abandoned the traditional objective of controlling geography in Vietnam. Battle lines moved by the hour. Landing zones (LZ) and firebases were won, abandoned, and then fought for again at great human expense.

The only metric remaining to quantify success was the enemy body count.

CHAPTER 4 PREPARING FOR WAR

I arrived at Fort Rucker in mid-April 1969. Command divided us into training classes.

At age 17, I was the youngest in my class. In a quirk of fate, I received my E-2 promotion (private first class) out of Basic Training a few days before anyone else in the 25 person class. That made me the ranking enlisted man in my class, so they appointed me Class Leader with the temporary rank of E-5 Sergeant (Non-commissioned officer).

It was a daunting experience.

Most of the guys were in their late teens and early twenties. Some were tough guys from rough neighborhoods. I knew from my experiences with unsavory people in Cushing and Wichita; I had to quickly garner respect.

On the first day, I called the Class into marching formation. Some older tough guys were slow to respond. At the top of my voice, with all the authority I could muster, I ordered the lagging soldiers into formation.

That seemed to do the trick. I never had a disobedience problem again during my tenure as Class Leader.

I had learned all the popular marching cadence songs that reinforced the drive to kill Viet Cong. So, as I commanded our unit in marching or running between classes and training sites, I led the cadence Kill songs.

I did not realize how that constant psychological reinforcement was transforming me and others into killing machines.

AIT was challenging. I seemed to have a knack for mechanical systems, so I did well. One of the most interesting tidbits I learned about Huey helicopters, there is one big nut that holds the entire main rotor blade system in place. If the nut comes off, the rotor system separates from the helicopter and everybody dies. The failure of the large retaining nut had such cataclysmic consequences, they nicknamed it the "Jesus Nut".

That bit of levity stuck with me and would save mine and my crew's lives in the not-so-distant future.

There are a lot of moving parts in a Huey helicopter and a lot of ways to be killed, notwithstanding enemy bullets.

A Lycoming turbine engine powers the helicopter. The engine drives an upright transmission that is connected to the main and tail rotor blades. It is imperative the pylon assembly, which is the mounting platform for the main rotor blades, and the tail rotor system is inspected daily.

The transmission has a free-wheeling clutch. When the engine powers down, inertia continues turning the rotor blades. This is a critical feature that allows the pilot to "auto rotate".

If the engine fails in flight, the pilot quickly flattens the blade pitch as the helicopter falls in an accelerating dive. As the ship nears the ground, the pilot pulls the collective lever up, which operates the main rotor blade pitch control rods.

The result is like blowing into fan blades. The helicopter blades are turned by the rushing air, creating enough lift to make a safe, yet harrowing, landing.

A dangerous hazard can develop if the helicopter is leaning on uneven ground or a hard landing damages the landing skids. The rotor blade arc can sweep low to the ground and fatally injure an unsuspecting crew member who does not get low enough when egressing the ship.

There is a twenty-minute and five-minute fuel light. If the twenty minute light illuminates, you need to start back to the POL dump (Petroleum Oil Lubricant) as soon as possible to refuel. If the five-minute light illuminates, you must LAND

NOW!

In combat conditions, if the enemy hits the fuel cell with bullets or shrapnel, the cell should be full. It seems counterintuitive, but full fuel cells tend not to explode. If you are low on fuel and get hit, you are at high risk of the fuel vapor in the cell exploding.

Helicopters burn fast. Manufacturers use magnesium in helicopter construction to reduce weight. When ignited, magnesium flames can reach 5,000 degrees Fahrenheit.

The Military Occupational Specialty code (MOS) for graduation from the Helicopter Maintenance School was 67N20. A 67N20 guy is usually the designated Crew Chief of an aircraft in flight platoons.

As I learned in AIT, few 67N20 graduates get to fly in the U.S. I had no intention of being a tool jockey in a maintenance hangar. I got into this man's Army to fly!

The aircraft is the Crew Chief's baby. The Crew Chief or CE for short (crew engineer) is the first line of authority regarding the aircraft's operational status. When the aircraft is on the ground, the CE is in command. The aircraft could not fly unless the CE's daily mechanical inspection was satisfactory.

If an issue is found that jeopardizes safe flight, the CE "red X's" the issue in the logbook, and the aircraft is grounded until the issue is repaired or replaced.

In Vietnam, many disregarded that regimen. It was common for crews to euphemistically "shake the stinger, if nothing fell off, fly it" (the stinger is a guard below the tail rotor to prevent ground strikes)

I think the common mindset may have been, "it doesn't matter–you are going to die, anyway."

The maintenance platoon performed extensive inspections every one hundred hours of flying time. Components that are at or exceed the TBO (Time Between Overhaul) are required to be overhauled or replaced.

Stinger 523 Lycoming Turbine Engine

Most flight crews in Vietnam completely relied on the maintenance platoon to do their job.

As the Helicopter Maintenance School wound down, we spent a day at door gunner school. It was more of an introductory class. Door Gunner 101. I flew in a Huey for the first time. The M-60 machine gun was mounted on a pylon with the trigger handle on the back of the weapon. We fired at stationary targets as we whizzed by at sixty knots.

The reality in Vietnam was quite different, but it was a good start.

I met a few Vietnam veterans while I was at Fort Rucker. I learned that if a helicopter maintenance guy wanted to fly in Vietnam, the best chance was to be assigned to an assault helicopter company.

The combat stories of the Huey troop-carrying helicopters, commonly called "slicks", inspired me.

The Huey gunship stories filled me with the wonder of glory, guts, and danger.

I was told only the best airmen get accepted to the gunships.

Huey gunships were much smaller than slicks. They carried no passengers or cargo. The gunships' only objective was to engage the enemy in mortal combat.

Stories of the gunship crews mesmerized me. They constantly risked death to save fellow infantrymen and helicopter slick crews. The written accounts spoke of the gun crews with reverence as they suffered high casualty rates.

Although being a gunship crew member was one of the most dangerous jobs in the Vietnam War, these brave volunteer warriors attacked the enemy with a vengeance.

The mythical heroes killed and faced death nearly every day they flew. As I became more enthralled with the prospect of flying in combat, I pestered my Vietnam vet acquaintances for more details. I wanted to know what the rules of engagement were. My training had prepared me to kill, so I needed to follow through on that process with soldiers who had been there.

It turned out none of my acquaintances had killed anybody. Only one guy was even in a flight platoon and he flew on slicks.

I learned that a small percentage of trained helicopter mechanics flew in Vietnam, and a much smaller percentage of aircrews flew in gunships.

The prospect of killing the enemy captivated me. Patriotic duty fueled my growing passion. I tried to visualize what it would be like. I needed to know if I had the courage to do my duty in extreme combat.

The knowledge I was seeking seemed to be unobtainable until I was actually in the war. Few soldiers I met had that level of experience.

The veterans, who had actual combat experience, were not willing to share. I realize now, combat traumatized most soldiers. They just couldn't talk about it.

I planted the seed. I wanted to strive to be a gunship crew chief/door gunner if I ever got to Vietnam.

I graduated from Helicopter Maintenance/Door Gunner School in mid-July 1969. I received the Crewman Wings, which was a big deal to me.

Vietnam would have to wait. My first duty station would be in Savannah, Georgia.

It worried me the war would be over before I could get there.

Aircraft Crewman Wings

CHAPTER 5 HUNTER ARMY AIRFIELD

The Army stationed me at Hunter Army Airfield in Savannah. My unit was a Huey helicopter company, part of the First Aircraft Maintenance Battalion. They promoted me to Specialist Fourth Class.

I reported to the First Sergeant and quickly asked about flight positions. The First Sergeant (commonly called "Top") told me there were only a few enlisted flight positions with no vacancies.

I wanted to avoid working in the maintenance hangar as a mechanic. Plan B–I told Top about my clerical experience and asked about a position in the Orderly Room. That worked. Top assigned me to the Orderly Room.

I enjoyed administrative work and excelled. I learned there was a significant advantage of being close to the center of power. I also became acquainted with the admin guys in personnel, finance, battalion, and brigade headquarters. The benefits were endless.

After a few months as Orderly Room Clerk, they awarded me the secondary MOS of 71B20 Clerk/Typist. The new designation would become a blessing and a curse.

Within the limits of my rank, I became friendly with the Commanding Officer (CO) who was a Captain, and the First Sergeant, who was an E-8 Master Sergeant.

After a few months, the CO invited me to fly with him on an early morning currency flight to maintain his flight status. I jumped at the opportunity. It would be my second flight in a Huey.

It was still dark when I met the CO and co-pilot on the flight line. After the preflight, the dawn was breaking. The weather was cool, but perfect. It thrilled me to hear the turbine engine winding up, which slowly started turning the rotor blades. The smell of burning JP4 jet fuel, first thing in the morning, was a fresh experience which I enjoyed very much.

I sat in the back of the cabin near the transmission well. The doors were closed. After take-off, the grinding sounds coming from the transmission surprised me. I thought something was wrong. I needed to avoid making a fool of myself by raising my concern with the CO. I reasoned that if there was an oil circulation problem or there was damage, the pressure and chip detector indicators in the cockpit would display the warnings.

I recognized my initial training was transforming me into an airman. I still had a long way to go.

I yearned for the chance to fly in combat. My duty as a clerk was a delaying tactic to keep me out of the maintenance hangar. I was becoming more concerned the war would end and it would trap me in my secondary MOS of Clerk/Typist. I was becoming desperate.

One day, the CO suggested I apply for Officer Candidate School (OCS). In the Vietnam era, OCS would accept High School GED applicants. So, that became my priority. I quickly passed the High School GED exam. High military test and intellect scores accompanied my application to OCS with glowing reference letters from my CO and the Battalion Commander.

As I pondered this new opportunity, I visualized attending flight school after OCS graduation and becoming a U.S. Army pilot.

They scheduled me for an appearance before the OCS Board, which would determine my qualifications to be an officer.

I had several weeks to prepare. I started an intense physical training routine. I studied all available reference material in areas they might question me.

About two weeks before the OCS Board appearance in late February 1970, I approached the CO and asked about the Board procedure. He answered my questions. I added, "Sir, I am interested in applying for flight school. Would I need to be commissioned in the Transportation Branch?"

He replied, "No, not necessarily. You might go to Infantry OCS. You would have the opportunity in OCS to take the flight school entrance exam. If you were one of the few that pass the test, you could go to flight school as an infantry second lieutenant instead of an enlisted candidate in the warrant officer program."

The big day arrived. Just a couple of hours before the life-changing event, I reviewed my appearance in the full-length mirror. My shoes were spit-shined to perfection. My Class A uniform was cleaned and pressed. My brass belt buckle and collar insignia were polished to a shining gleam. I had a fresh short military haircut. I closely shaved what peach hair I had for whiskers. Everything was perfect.

I drove to Brigade Headquarters. There was no wait. I was the first to be interviewed. I marched in and stood at attention before the ranking officer. Two captains and three field-grade officers comprised the panel.

I saluted and stated, "Sir, Specialist Fourth Class Allen reporting."

The full bird colonel returned my salute and commanded at ease. The questioning was intense. My preparation paid off. The interview lasted for about twenty minutes. The Colonel dismissed me.

About seven days later, the CO called me into his office. He said, "Allen, we got the OCS Board decision. Congratulations, they have accepted you to Infantry Officer Candidate School in Fort Benning, Georgia."

April 20, 1970 - *President Nixon pledged to withdraw 150,000 more troops from Vietnam over the next year and once again appealed to the North Vietnamese to undertake serious negotiations.*

April 30, 1970 - *President Nixon announces the Cambodian Incursion - In a nationally televised address, President Nixon announced he had sent 2,000 American combat troops into Cambodia and ordered U.S. B-52 bombers to begin airstrikes. Nixon reversed his April 20 announcement that he would withdraw 150,000 troops from Vietnam over the next year, in effect providing that there would again be a need to draft young American men to maintain the current force level. Nixon told viewers, "This is not an invasion of Cambodia. The attacks were upon the territory in Cambodia that was "completely occupied and controlled by North Vietnamese forces."*

May 2, 1970 - *Violent anti-war rallies were staged at many universities.*

May 4, 1970 - *Students at Kent State University in Ohio conducted a protest against the Vietnam War. The National Guard responded. Four students were shot and killed. The event is often cited as a turning point in the war.*

May 8, 1970 - *Jim Cairns, a member of the Australian parliament, led over 100,000 people in a demonstration in Melbourne. Smaller protests were also held on the same day in every state capital of Australia. Australia was an American ally and had troops in Vietnam. 60,000 Australians served, 521 were killed, and over 3,000 wounded.*

May 9, 1970 - *A crowd of about 100,000 demonstrated peacefully in the event organized by the New Mobilization Committee to End the War in Vietnam in one of many anti-war protests that took place* *across the nation that day. Even though the demonstration was quickly put together, protesters were still able to bring out thousands to march in the Capital. It was an almost spontaneous response to the events of the previous week. Police ringed the White House with buses to block the demonstrators from getting too close to the executive mansion.*

May 14, 1970–*Tennessee Jackson State College - Two dead and twelve injured during violent protests.*

May 20, 1970 - *New York. An estimated 60,000 to 150,000 are at a pro-war demonstration on Wall Street.*

May 28, 1970 - *University of Tennessee, Knoxville. Nixon at Billy Graham Crusade in Neyland Stadium - 800 Students carry "Thou Shalt Not Kill" signs into the stadium. Many are arrested and charged with, "disrupting a religious service". Only Republican candidates were on the stage with Graham and Nixon.*

CHAPTER 6 OFFICER CANDIDATE SCHOOL

I reported to OCS on March 29, 1970. Everything else I had experienced in the Army, which I thought was difficult, was a walk in the park compared to OCS.

I was the second youngest in my class of over 130 candidates.

OCS is Basic Training on steroids. Candidates were pushed far beyond what they thought was their physical limit. The constant running every day on asphalt wearing combat boots gave us "OCS knees". Many suffered painful knees from battered joints.

We had to do at least ten front-handed pull-ups without swinging on the ten-foot-high pull-up bars that were constructed at the entrance of the mess hall.

If you failed to properly complete the pull-ups, you could not enter the mess hall to eat. The number of required pull-ups kept increasing.

OCS stoked the patriotic fire and passion to kill the enemy.

I was doing fine with the academic studies and coping with the physical training. I had the genetic intellect to succeed. I did well as a very young Class Leader in AIT. So, I had some leadership ability.

The psychological torture became the crucible for me. I had lost the much-needed life guidance that a father provides his son. I had done alright to that point, but I lacked the foresight I desperately needed to make it through the rigorous training.

After a few weeks, they invited candidates to apply for flight school.

Several applied including me. Because of my previous studies in aviation, I could function at a higher level in that subject. I was one of two candidates that passed the exam.

Severe harassment for sixteen hours a day was a reality I was ill-prepared for. I needed to focus on getting through one day at a time. I had to see beyond the remaining few weeks in the 13-week course to graduate.

President Nixon's April 30 announcement that America would expand the war into Cambodia increased the demand for new Army infantry officers.

Another consequence of the President's announcement was most OCS graduates and non–graduates received orders for Vietnam after OCS.

My acceptance into OCS created the opportunity I had sought for many months. They would assign me to duty in Vietnam.

If you could make it to the weekend, you would get some time off to recoup. Alas, if candidates showed a lack of resolve, the instructors would assess demerits for insignificant issues, causing the loss of the weekend reprieve. The beleaguered candidate would be forced to march "tours" back and forth in a parking lot for the entire weekend.

I caught a day of cold driving rain to walk tours on my second infraction. That was the proverbial straw that broke my back. I resigned from my candidacy.

My bridge had not been burned. I had my aircraft crewman's wings. I was qualified to fly as a crew chief/door gunner. I was eager to go to Vietnam as a trained helicopter crew chief. Maybe, if I was lucky, I would be accepted into a Huey gunship platoon.

In retrospect, OCS was very beneficial for me. It made me a much better soldier.

The skills I learned served me well for the duration of my military service.

These attributes proved to be priceless in civilian life.

As I reflect on that period as an eighteen-year-old, I realize I had a diminutive understanding of what it meant to be a commissioned officer in the United States Armed Forces.

With no higher formal education, troubled adolescence, and no advanced parental guidance, I failed to understand the scope and expectations of an officer.

As war events continued to unfold, it would become apparent my early desire to make the Army a career was premature.

CHAPTER 7 GOOD MORNING VIETNAM

After my thirty-day leave at home, my orders specified reporting to the Oakland Army Terminal in California for deployment to Tan Son Nhut Airbase, Saigon.

Before I departed for my leave, I had to report for the Vietnam vaccine protocol. The medic brought out a tray of about twenty syringes and vials. I remarked, "Oh, I thought I was the only one processing today." The medic chuckled and said, "You are. These are all for you."

I spent thirty days at home with my girlfriend and family. After a couple of weeks, I became restless. It was harvest season in Kansas. Farmers needed combine drivers to cut wheat. My experience on the custom harvesting crew just two years prior was still fresh.

A local farmer hired me, and I worked for him for about ten days. I got to know him and his family. They were neutral about me going to Vietnam to fight. They clearly did not support the war. The family wished me well and prayed for my safe return.

One night, at my parent's home, I was up late watching TV, alone. The broadcast day ended. The custom was for the TV station to play the National Anthem with heart-soaring videos of America's beautiful plains, mountains, beaches, gleaming cities, and rustic small towns. I surprised myself as the broadcast sign-off compelled me to stand at attention and salute the image of the American flag waving in the wind.

My patriotic pride was intense.

After the leave, I started my journey in late June 1970. Traveling in Class A uniform evoked different responses from other passengers during that time.

Most people were indifferent. Some would react negatively. No one thanked you for your service.

The anti-war protests in Oakland had petered out, so I was not exposed to that when I arrived at the Army Terminal.

It was afternoon when we began checking in. We went through the processing and issuance of jungle fatigues and gear.

In the processing area, returning troops from Vietnam were in a short line on one side of the area. The troops being deployed were about 20 feet away in a long line.

The contrast in demeanor was shocking. All the "new guys" were excited, joking around, and expectant of glorious adventures ahead. We all looked like new guys in our brand new, crisp jungle fatigues and boots.

The returning veterans looked sullen and exhausted. Their uniforms were worn and faded.

I will never forget when I broke from the frivolity occupying our line and caught one of the returning guys looking at me with a sad gaze.

The war-weary soldiers did not kid or prod the new guys as we passed in two different directions. They only projected sadness. They knew what we were going into.

We did not know what Vietnam had in store for us.

Our long flight was on a chartered civilian airliner with congenial flight attendants. There were no noticeable events. Just a lot of napping and anxiety.

Unlike other wars, they deployed Vietnam soldiers one at a time and came home one at a time, if they lived. We were not trained and deployed as units. So there was no sense of continued camaraderie.

A tour in Vietnam was a very finite one-year experience a soldier endured alone, except for the bonds formed in-country.

It was early morning as we approached Tan Son Nhut Airbase near Saigon. Excitement and trepidation were building. The thought crossed my mind. The enemy could shoot us when we disembarked the aircraft.

I did not know where the dangerous areas were or if everywhere was dangerous.

We landed and disembarked without incident. After about an hour's wait, we were shepherded into buses for a trip to Long Binh, about ten miles away.

The heavy wire mesh on every passenger window of the military bus was disconcerting. I surmised it was a defense against the enemy throwing grenades into the bus.

I still had not been issued a weapon. I felt vulnerable.

The wire mesh on the bus windows deflated our belief we were liberators. No cheering crowds like Europe during World War II, welcoming the forces of freedom. Instead, we got defensive measures to protect us from the populous so they did not kill us before we were even issued a weapon.

Upon arrival at Long Binh, they assigned our barracks, and we got some chow. Then the wait started again. We were not told how the entire process worked to be assigned to a unit. The wait lasted a few days. The unknown was torturous.

I finally got the call to report to First Aviation Brigade processing. I arrived at Brigade in the early morning and started another waiting period. After a few hours, they called my name.

I sat down with an Army Specialist Fifth Class admin guy. He reviewed my file and noticed my 71B20 Clerk/Typist secondary MOS.

He stopped, looked at me, leaned back, and said, "You don't want to go to an aviation unit, do you? I can assign you to an admin job right here in Brigade. We need clerks. It is like stateside duty. It is safe. We often go into Saigon, experience the ladies and nightlife. Life is good here. They measure life expectancy for aviation combat in minutes. The body bags stream in every day from the front line aviation units."

He looked at me straight in the eyes and said, "What will it be?"

I was shocked.

It never occurred to me, I would have choices. My focus was on serving in aviation. But my concept of aviation in Vietnam was not clear. The immensity of danger had never been suggested to me in specific terms.

The glory of war dazzled me.

After a brief pause, I slightly stuttered, "I want to serve in a combat assault helicopter company."

The admin guy looked incredulous, shrugged his shoulders, leaned forward to his desk, and dutifully assigned me to the 25th Aviation Battalion in Cu Chi. He handed the paperwork to me like it was a death sentence.

In a low voice, he wished me good luck.

I had to come back the next day to pick up my formal orders. I exchanged my U.S. dollars for the equivalent military payment certificates. I then arranged for transportation to the 25th.

It was about an hour and a half ride northwest to the 25th Aviation Battalion in Cu Chi. I was becoming familiar with the surroundings. It was extremely hot and humid. The smells were different. Most women wore silk tops and bottoms with large conical leaf hats. Different variations of motorcycles were the primary vehicle with two-cycle engines belching out smoke and fumes. Everyone else was on bicycles or in unfamiliar automobiles.

No one noticed another soldier making his way through their

country, except several vendors.

One vendor offer, in particular, puzzled me. "G.I. wanna boom-boom?" I got that one a few times. I later learned it was the Vietnamese version of pimping girls.

I was not in Kansas anymore.

It perplexed me. Here I am in a war zone. I have not received a weapon yet. Where is the enemy? How does the enemy look any different from the people I was seeing?

Upon arrival at the 25th Aviation Battalion, we offloaded, shouldered our stuffed duffel bags, and entered another processing waiting area. This time, the wait was minimal.

I sat down with a buck sergeant (E-5) admin processor. He looked at my orders, did a cursory review of my records, and stopped. The sergeant happily stated, "Oh great, you are a clerk/typist. We desperately need clerks. We are going to keep you here. Welcome to Battalion Processing."

I became desperate. This is not what I wanted. I had to fight.

I regained my senses and protested. "Look, I need to be assigned to a combat assault unit using my primary MOS. I want to fly."

The Sergeant sternly replied, "Specialist Allen, you will go where the Army tells you to go."

I could feel my dream slipping away.

My experience in dealing with Army command and bureaucracy kicked in.

I retorted, "Sergeant, I request to see your superior."

Sensing he was getting into a pissing match that might not go well for him, the Sergeant regrouped and became spiteful. He shot back, "Oh, OK, you are the John Wayne type. They get John Wayne types going through graves registration (morgue) every day."

The processor continued, "Alright, the 116th Assault Helicopter has a couple of open slots. We will assign you to them, uh Mr. Wayne.", he said with a snicker.

June 6, 1970 - *At the commencement of the University of Massachusetts, students stenciled red fists of protests, white peace symbols, and blue doves onto their black gowns.*

June 24, 1970 - *Congress repeals the Gulf of Tonkin Resolution to reassert control over the president's ability to use force in the war. The Resolution had supported U.S. intervention in the Vietnam War since 1964.*

CHAPTER 8 THE 116TH ASSAULT HELICOPTER COMPANY

It was a quick ride by jeep to the 116th company area. The driver pulled up near the orderly room.

I retrieved my duffel bag from the back of the jeep and walked across a small footbridge that spanned an awful smelling ditch. I noticed there were several latrines (outhouses) nearby. I would later become too familiar with that ditch.

As I walked through the orderly room door, I saw a couple of clerks.

I asked the nearest guy if I could see the First Sergeant. He pointed to a Master Sergeant sitting at a desk across the room. I walked over and blurted out, "Good Afternoon First Sergeant. I am Specialist Fred Allen. Battalion has assigned me to the 116th AHC. Here are my orders."

Assuming he had not received my records from Battalion, I immediately made my pitch. "First Sergeant, my primary MOS is 67N20. I am qualified as an Expert in M-14, M-16, and M-60 machine guns. I request to be assigned to the gunships, if possible."

The orderly room fell silent.

After an uncomfortable pause, the First Sergeant responded, clearly agitated by my insolence. "Well, Specialist Allen, where you will be assigned, is up to the Commanding Officer. New enlisted men are not usually assigned to the guns. You would need some experience in slicks first."

Secretly, I was relieved to hear at least I had a chance to fly.

The First Sergeant continued, "You have to report to the Commanding Officer and he won't be back for another hour."

I stowed my duffel bag and asked for directions to the mess hall.

I walked through the company area to the mess hall for chow.

I stuck out like a sore thumb in my BNFNG fatigues (Brand New Fucking New Guy).

A BNFNG had no status.

You were just a blob of protoplasm taking up someone else's space.

In Vietnam, with a one-year tour of duty, you had to get up to speed fast in combat units. New guys could get a seasoned guy killed.

There was a time progression scale on your status. You were a BNFNG for a couple of weeks. Then, for about two to three months, you were an FNG. After that, if you were still alive, you attained full status.

The FNG period could shorten if you quickly proved yourself in combat.

Your combat awareness followed a similar progression. In the beginning, you are too stupid and inexperienced to be scared.

Training might prepare you at a conscious level, but only experience can alter you at the instinctive level. After you get shot at, witness death, or endure incoming rockets, it sinks in. Just being in Vietnam was hazardous.

After the break-in period, for the next several months, you realize you are a dead man walking. There was no way you were going to survive assault helicopter missions and the occasional incoming rocket attacks.

So, for those interim months, nothing mattered, because you were already dead with no hope and no future. It is during this period, a combat soldier is most likely to commit the most heinous or heroic acts. He is not scared of anything; he is already dead. He is just angry and wants to take as many of the enemy with him as he can.

The last month or two were the worse. You were short.

The specter of hope that you might survive and the fear of dying was increasing. The shorter you got, the more hopeful you became, and the more scared you got.

You thought about life after Vietnam.

I returned to the orderly room an hour later. The First Sergeant escorted me to the Commanding Officer's office. I marched into the room, stationed myself in front of the CO's desk, snapped to attention, and saluted. "Sir, Specialist Fourth Class Allen reporting for duty".

The CO returned my salute. I lowered my salute. The CO commanded, "at ease". I relaxed into a parade rest stance.

The CO was a Major, which I was not accustomed to. All the company-level COs I had encountered before were Captains or even First Lieutenants. In the States, they usually assigned Majors to a battalion or higher. But this was not a training or support unit. This was the real thing. Millions of dollars of assets and soldiers' lives were at risk.

The CO started, "Allen, we got your records from Battalion. I have spoken to Lieutenant Fisher. He is the Stinger gunship Platoon Leader. He is waiting to see you. Dismissed."

I snapped to attention and saluted. The CO returned my salute. I pivoted 180 degrees and marched out of the room.

I contained my exuberance while I asked for directions to Lieutenant Fisher's quarters, or "hooch," as they were called.

It appeared I was going to the guns!

It was a level of excitement I had never felt.

I was joining this mythical group of air warriors!

I retrieved my duffel bag, slung it over my shoulder, and with grand purpose, briskly walked over to Lt. Fisher's hooch.

I knocked on the screen door.

A young, battle-worn First Lieutenant appeared. The name Fisher was emblazoned on his faded flight suit.

I saluted, "Sir, Specialist Allen reporting as ordered by the CO."

The Lieutenant returned the salute and opened the door. He did not invite me in.

As we stood at the doorway of the plyboard and screen "hooch" the Lieutenant started, "Allen, are you familiar with the Stinger mission?"

I answered, "Generally, sir."

The Lieutenant continued, "We have two primary mission objectives. Provide protective cover for the slicks and grunts on the ground. Second, search and kill missions. The Platoon Sergeant is Sgt. McKinley. He will fill you in on the details."

Lt. Fisher added, "We usually only accept exceptional crew chief/door gunners after they have flown with the slicks for a while or transferred in with experience."

The Lieutenant seemed to become slightly irritated. "We are making an exception in your case. The CO thinks that since you had what it took to be accepted to Officer Candidate School, you may have the grit to be a Stinger."

Lt. Fisher's tone changed to subtle indignation. "I respectfully disagree with the CO's viewpoint, since you quit OCS. But it is the CO's decision."

The Lt. sighed, "The second reason is we are short. The enemy killed one of our crew chiefs yesterday."

The Lieutenant's face turned to stone as he issued a warning. "Allen, duty in the Stingers is extremely dangerous. The crew chief's performance often means the difference between life and death for the entire crew. We will evaluate you daily until we are confident in your abilities."

I stood at attention, saluted, and responded, "Yes Sir".

I asked for directions to Sgt. McKinley's hooch.

Sgt. Curtis McKinley was a tall young man in his early twenties. He had an air of wisdom beyond his years, shrouded in a no-nonsense demeanor.

The Sergeant extended his hand, and we shook.

McKinley began, "Allen, we rarely get inexperienced guys, so we have to get you up to speed fast. We have a mission tomorrow morning at 0800, so it will be a 0630 wake-up, 0700 breakfast, 0730 flight line, and 0800 lift-off. I am assigning you aircraft 523. Your door gunner is SP4 Bremer. Bremer can get you up to speed on the armament. I believe it is a search and kill mission. We will go back to the area where your predecessor was killed yesterday."

The Sergeant finished, "Any questions?"

I had plenty of questions. I just didn't know how to articulate them.

I managed one question no one had ever given me a straight answer for. "What are the rules of engagement?"

McKinley looked at me strangely for a moment and then said, "If they are shooting at you, kill them. If they run, kill them. If they have any military gear, kill them. If you are not sure, kill them."

McKinley continued, “After you get settled, we need to go to Supply so you can draw your flight gear and personal weapon.” The Sergeant took me to my hooch, which was shared with thirteen other men. McKinley pointed out my bed and left me to unpack.

No one else was there. I sat down on my bed and just tried to gather my thoughts. I recalled being back at AIT in Fort Rucker and making the pledge to become a gunship crew chief/door gunner if I ever got to Vietnam. I could have never imagined the path I would take to make that pledge a reality.

Clearly, without my acceptance into OCS, I probably would not have been assigned to Vietnam. I certainly would not have been accepted into the Stingers as a BNFNG.

There was an empty footlocker at the end of my bed with the name Rowland inscribed on a strip of masking tape. A small photograph of two young girls was lying on the floor. My bed and small storage area may have belonged to the crew chief who was killed the day before. The officer in charge must have missed the picture of the two little girls when he gathered the personal effects.

A couple of days prior, I escaped being chained to a desk as a clerk/typist at Brigade, then Battalion. Now, tomorrow morning at 0630, I will get up. Go to the flight line after breakfast. Get familiarized with an aircraft I had never seen and do the preflight.

Quickly, learn the armament system from SP4 Bremer. Then strap in and take off on a search and kill mission where my predecessor was killed two days prior.

For the first time, I might kill...... or be killed.

After getting squared away, I looked around the hooch. I saw a small American flag displayed on a wall next to two rectangular helicopter inspection plates. The plates were reversed to expose the unpainted side. There was a lot of felt pen writing on both plates.

One plate had a heading "KIA" (Killed in Action) and the other plate heading was "Lost Aircraft". The KIA listing had 26 enlisted men's names and the Lost Aircraft plate had 17 aircraft tail numbers. It appeared to be the enlisted Stinger memorial monument.

Apparently, the officers kept their own KIA tally.

I remember the mental impact of seeing the cost in lives and aircraft for just this one platoon of gunships. A heavier weight of reality settled upon my psyche.

McKinley's response to my question about the rules of engagement is the most profound edict I have ever encountered.

Simply, it was the standard operating procedure to kill any Vietnamese in the countryside, while in flight, at my discretion.

The Vietcong (VC) were guerilla fighters. Most VC were local villagers who were recruited, provided minimal military training, and supplied with basic weapons. The VC did not have standard uniforms. They were clad like any villager or farmer.

The VC used the geography of the land to strike and hide. They used vast tunnel complexes as cover from U.S. airstrikes and large-scale ground attacks.

In III Corps around Saigon, usually, only VC were engaged.

I learned there was another class of enemy. The North Vietnam Army (NVA) uniformed soldiers were better trained, equipped, and motivated than the VC.

In I Corps, the northernmost region of South Vietnam, the NVA was the prevalent enemy.

McKinley returned about an hour later. We walked over to Supply. They issued me a flight suit, gloves, flight helmet with a microphone, and an M-16 with ammo. The flight suit had the Stinger platoon patch on the left chest.

As I was signing for the gear, the supply sergeant said, "sorry, we are out of chicken plates".

I was not familiar with the term "chicken plate" and I did not want to emphasize my ignorance by asking.

From my previous conversations in the States and reading pertinent literature regarding Vietnam's flight platoons, I could piece together how they worked.

In Hueys, there was a four-man crew: Aircraft Commander (AC), Co-Pilot (CP), Crew Chief (CE "Crew Engineer"), and second Door Gunner (DG).

The Aircraft Commander was in charge of the aircraft in the air and did most of the flying. The Co-Pilot backed up the AC. The Co-Pilot may or may not outrank the AC. Rank was not an issue in flight. The AC was in command of the ship. The Co-Pilot position was a necessary step to becoming an Aircraft Commander.

The Crew Chief had two major responsibilities. He was in charge of the airworthiness of the aircraft and was in command of the aircraft on the ground. In the air, the Crew Chief was the primary Door Gunner.

The second Door Gunner often had no aviation training. Volunteer infantry guys with M-60 experience could transfer in to take Door Gunner positions. The second Door Gunner was in charge of the ship's armament.

I made my way back to the hooch. Some guys had returned. I shook hands. Everyone was friendly in a subdued way.

I heard Jerry Watson, a CE from California, say, "Hey guys, when will the fun start", slyly looking my way.

That was ominous.

There were some chuckles and snickers. Everyone continued what they were doing. The mood was somber. I assumed it was because of losing the crew chief the day before.

After a while, I walked over to Tony Bremer's bed.

Bremer was an SP4 Door Gunner from Alabama. After a little chitchat, I stated, "It is my understanding we will fly together tomorrow." Bremer responded with a pronounced southern drawl, "Oh yeah, news to me."

He was not thrilled about flying with a BNFNG crew chief.

Undeterred by the lack of congeniality, I pursued, "do you mind if we talk a little."

Bremer half-heartedly responded, "Fine with me. Sit yourself down."

I sat on the opposite bed. "Can you tell me about the ship's armament?"

Bremer replied, "Sure. I can do that, but you must go through an actual hands-on orientation before the flight tomorrow."

I replied, "OK, will do."

Bremer continued, "We have two M-60 machine guns and two 19 hole rocket pods for 2.75 rockets. We carry an M-79 grenade launcher with about thirty grenades. We have an ammo can full of different hand grenades: frag, incendiary, white phosphorus, CS Gas, yellow, red, and purple smoke. I carry an M-14 and an AK-47 I got from a grunt. The 2.75 rocket warhead will vary depending on the mission. Usually, we use the 17 pound High Explosive, which is as effective as a 105 mm Howitzer. Sometimes, we use the 10-pound Flechette warhead, which releases about 1,000 small arrow-shaped projectiles on impact."

I responded sarcastically, "Wow, no nuclear weapon?"

Bremer slightly smiled, "we are workin' on that."

Bremer suddenly remembered, "Oh yeah, I have a grease gun".

It puzzled me, "What is a grease gun?"

Bremer described, "A grease gun looks like a handheld lubricant grease gun that automatically shoots 45 caliber bullets."

I sat back, amazed. "45 caliber rounds, holy shit. That would knock a VC down, but you would have to be close."

Bremer responded, "Right–it is a weapon of last resort if you crash and survive."

Bremer added, "Allen, just to be straight, we don't call 'em VC, NVA, or anything else that might dignify them. To us, they are just dinks."

I cautiously asked, "What was the crew chief's name that was killed yesterday."

Bremer's demeanor changed. His voice became strained. "His name was Mark Rowland. He was an SP5 with one month left in his tour. He was a great crew chief with over forty personal kills. Everyone liked him. He was married with two young daughters. I was the door gunner with Mark when he got hit."

I paused for a few minutes. Bremer was a little emotional after talking about Rowland.

I continued, "McKinley told me we might go back to the same area where Rowland was killed."

Bremer responded, "Good, I'll be hope'n we get payback."

Later, I saw Bremer walk over to the Stinger Monument and write in "SP5 Mark Rowland" on the KIA plate.

Conversation in the Stinger hooch momentarily paused in silence.

When 1700 hours rolled around, everyone started for the mess hall. I had changed into my flight suit to distance myself from the BNFNG look.

Almost as an afterthought, I placed my pocket-size Bible in my front pocket. I was not a devoutly religious person, but I felt like I might need all the help I could get.

As we walked into the mess hall, there were about thirty other guys already there wearing flight suits adorned with platoon patches for the slick units; Yellow Jackets, and Wasps.

As we walked through the chow line and sat down at a table, I noticed the other platoons treating the Stingers with some deference.

As time went on, I understood the difference between slicks and gunships.

The slicks were often in harm's way. They presented a ripe target for the enemy when they were approaching or departing a landing zone. They were sitting ducks when loading or unloading troops and supplies in unsecured areas.

The slick door gunners had a limited range of fire, with M-60s mounted on pylons. They also had other ships and grunts on the ground, obstructing their field of fire.

Slick guys recognized it took a different breed of airman to be in gunships. You had to have a killer mentality. You had to relish killing another human every time the opportunity arose.

The gun crews were not sitting targets in a hot landing zone taking intense enemy fire.

The gunships were circling above the enemy, doling out death and destruction.

Slicks were in the business of transporting.

Gunships were in the business of killing people.

After supper, we straggled back to the Stinger hooch. About an hour later, Lt. Fisher came in and announced a memorial service for SP5 Mark Rowland would be held at the mess hall tomorrow at 1900 hours.

As the Lieutenant was leaving, he turned and asked, "Has Allen been initiated? If not, he is flying tomorrow, so it has got to be done tonight."

I had stripped down to boxer shorts and a t-shirt to prepare for sleep, hoping to get a good night's rest before a big day tomorrow. The Lieutenant's initiation announcement was a complete surprise.

The guys were complacent at first, still mulling the memorial service, then McKinley said, "Fuck it, let's do it." The supporters rallied, encircled my bunk, picked me up, and started carrying me toward that awful smelling ditch.

As I was struggling to free myself from my captors, Watson, the CE from California, was shouting in my ear, giving me details, "Allen, all of those latrines drain into the ditch. That is why we call it the shit ditch. To be a Stinger, we must throw you into the shit ditch."

Onlookers gathered along the way to laugh, jeer, and cheer the infamous Stinger initiation. The procession arrived at the bank of the shit ditch. The odoriferous aroma filled the evening's warm, moist air. I was unceremoniously tossed into the bacteria-laden soup. A unified cheer rose from the Stingers and onlookers alike.

As I tried to climb out of the cesspool, everyone was walking away, offering me no help. The banks were slick and slimy. I had to struggle through several feet of unimaginable filth to find an area of the bank I could climb. I finally made it.

I did not know where the showers were. I had to stop and ask someone who was repelled by my aroma.

There was no light inside the shower. I peeled away my boxer shorts and T-shirt. A barrel situated atop the stall-like structure provided water. I fumbled around until I found a rope hanging from the top. I pulled it and was shocked by a torrent of cold water.

I let out a blood-curdling scream.

I scrubbed every inch and orifice of my body for as long as I could in the cold water with no soap. I rinsed my boxer shorts and t-shirt. I gingerly put the cold, wet clothes back on.

I walked into the Stinger hooch, dripping wet, which evoked another round of laughter and jeering.

As I dried off and put on fresh dry boxers and a t-shirt, Bremer came over and shook my hand, smiled, and said, "Welcome to the Stingers"

Several other guys did the same.

I later learned I had a few more hazing rituals to endure. I suspected I was getting preferential hazing treatment. Lt. Fisher probably told McKinley to give me the full course to shake me out of the Stingers.

I was also told, if I passed the initiation ritual, I would have one more requirement to meet to become a full-fledged member of the Stingers.

Initiations and hazing are, of course, common in society. The severity of the mistreatment varies. I suppose it is commensurate with the function of the group or club. The universal aim is to find out if the newbie has the necessary "right stuff" to be a member of the club.

As a recent OCS candidate, the Stinger hazing was minor compared to what OCS dished out. So, I sailed through the Stinger harassment without breaking a sweat.

The hazing continued for the next two days. I endured my bed being short-sheeted (you pull back the top sheet to crawl in only to be stopped because the bottom sheet is half folded).

They put a snake under the cover of my bed. The snake was as surprised as I was when I pulled back the cover to climb in.

They threw me into an empty clothes locker and locked the door. Lighter fluid was then sprayed inside the vent and set on fire.

The last act was being hung up with handcuffs to the hooch rafter while I stood on my bed for about an hour.

One of the Stingers walked in and asked, "Allen, what are you doing?"

I responded good-naturedly, "Just hanging around."

I was released a few minutes later.

That was the end of the initiation.

I just had one more requirement to meet.

I surmised the last prerequisite was the most important.

Did I have what it took to kill an enemy soldier in combat?

The Stinger Platoon Patch

CHAPTER 9 MY FIRST MISSION

The Hornets had been part of the Cambodian invasion announced by President Nixon on April 30, 1970, with major assaults in the "Parrot's Beak" (south of Tay Ninh).

There were intense encounters with the NVA and Viet Cong.

A Stinger ship was hit and crashed. The entire crew was killed.

The Yellow Jackets and Wasps also suffered losses.

I missed that by a couple of months and could get broken in with missions in the Cu Chi tunnel complex areas where intense encounters were more sporadic.

As several of us walked to breakfast that morning, I smelled an odor that was so repugnant I forgot how hungry I was. I squeezed my nose to block out the stench.

I rhetorically asked, "What in the HELL is that smell? "

Everyone laughed.

Bremer said, "Welcome to Vietnam."

I was told the noxious smell came from the Army's method for disposing human waste. The latrines had 55-gallon barrels that were cut in half and served as receptacles for feces. A local Vietnamese worker would drag these foul-smelling containers out from under the latrines and pour in diesel fuel. The mixture was then set on fire to eradicate the waste.

Probably the most common memory for most Vietnam veterans is the aroma of burning shit and diesel fuel, first thing in the morning.

After breakfast, we all climbed aboard two small trucks and headed for the flight line. Upon arrival, McKinley took me aside and updated me on a change of plans.

I would fly with him on his ship, 470. I would be the right-side door gunner. They decided I needed some indoctrination during a live mission. McKinley would instruct me on the finer points of Stinger combat operations.

It takes a certain breed of soldier to climb into a gunship every day, knowing there will always be enemy contact. Every day a Stinger flew, he got shot at.

I completed the cursory inspection of the guns and rockets without Bremer's full orientation. I then asked McKinley about the aircraft inspection. He said the pilots usually took care of that. That struck me as odd.

As we stood by the ship's tail boom, McKinley went through the air procedures. "Stingers fly with the aircraft commander and crew chief sitting on the left side. We make left-hand orbits around a target. So the crew chief gets most of the M-60 kills."

A maintenance tech approached and interrupted McKinley with a couple of questions regarding the aircraft.

I jumped at the chance to look over the legendary gunship I had heard so many stories about.

It was an old ship with several sheet metal repairs in the fuselage to patch bullet holes. I was a little shaken to see bullet holes in the deck near the gunner's seats. I got a grip on myself to maintain my resolve.

They did not patch these holes. The deck was made with heavy gauge metal. They pounded the upturned rough edges of the bullet holes down with a hammer.

I could see there were no pylons in the gunship for mounting the M-60s. The gunners held a stripped-down version of the M-60 on their laps. This enabled the gunners to have complete maneuverability when engaging the enemy.

McKinley finished with the tech and resumed his orientation.

"A few ACs are extremely accurate with the rockets. The best AC with rocket accuracy is CW2 Lou Baldassaro. He is trying to get to 100 kills."

McKinley continued, "After the AC starts the turbine engine and is ready to lift off, the gunners visually clear the areas around the ship and holler, Clear Left and Clear Right."

The Sergeant added, "The CE pushes the side panel of the armored AC seat forward. The seat is designed in a way that makes it difficult for the pilot to slide the panel forward."

McKinley wrapped it up, "After crossing the perimeter of the base camp, the co-pilot pushes in all rocket-firing circuit breakers in the overhead panel. Both gunners would then lock and load their weapons and release safeties."

My first mission would be my third flight in a helicopter of any kind. The cargo doors in Stinger gunships were secured open or removed. Unlike the pilot's protective armored seats, the gunners sat on nylon-covered jump seats with no built-in armor.

Nothing prevented gunners from falling out of the ship during steep turns except a loose seat belt. Some ships had "monkey" straps for the gunners, which were about six feet long and fastened to a cargo floor attachment.

The right-side door gunner seat in 470 had an armored chest plate attached to the bottom of the seat.

McKinley said, "That's a chicken plate. It's designed for the chest, but most guys secure it to the bottom of the seat to protect their family jewels."

That kind of made sense. Most guys were around nineteen years old. They would rather be dead than get their genitals shot off.

I responded, "I hope my ship, 523, has a chicken plate under the crew chief seat."

McKinley dourly replied, "Yeah, it's got a chicken plate under the seat. 523 was your predecessor's ship, Rowland. He took a bullet to the chest, so the seat chicken plate did not do him much good."

The supply sergeant had already told me he was out of chicken plates, so no chest chicken plate for me. As time went on, the chest chicken plate lost significance. You were going to die one way or another.

As my mission experience grew, I learned to loosen my seat belt with enough room for me to maneuver my field of fire.

There were some exhilarating moments when I was leaning out of the helicopter firing the M-60, bracing into the recoil, then abruptly experience a jam.

The result would be the start of my fall out of the helicopter. That loose-fitting seat belt attached to the aluminum tube frame jump seat prevented my probable demise.

We lifted off and crossed the base perimeter. Rockets and M-60s were armed and ready to go.

It thrilled me to experience a low-level flight on that first mission. It was exhilarating to sit in the open doorway, half in and half out of the ship, watching the top of the jungle and intermittent rice paddies pass by. It had not soaked into my mind yet. People out there wanted to kill me.

I was a nice, exposed target in the doorway of that gunship.

As we entered the area where Rowland was killed, McKinley said, "OK, keep your eyes open and if you see tracers or hear gunfire, get on the intercom and yell, TAKING FIRE. The AC will bank the ship into left-hand orbits around the target. I will focus on the target and you provide cover on the right side of the ship."

McKinley added, "If the AC lines up to fire rockets, make sure your foot is not dangling in front of the rocket pod."

It surprised me to see the AC take a grease pencil and mark an aiming cross on the cockpit front windscreen for firing rockets. There was a rocket sighting device mounted on the right side where the pilot-in-command normally sat. Stinger ACs preferred the rudimentary aiming cross as they aimed the entire ship at the target to deliver the 2.75 rockets.

The left-side pilot seat provided better visibility.

A few minutes later, I heard McKinley scream on the intercom, "I got a dink running on the left, bank left."

The AC pulled the ship into a steep left turn and got a fix on the dink. McKinley was about to fire when "crack-crack-crack" was heard above the ship and wind noise.

Instantly, McKinley yelled, "Taking fire" and then opened fire with his M-60.

My sense of acuity exploded as I leaned way out of the right side of the ship to see while we were in a steep left bank.

I scoured the jungle, looking for movement. I thought I saw something and started firing.

Abruptly, the ship righted, and we continued the patrol.

I keyed the mic and asked McKinley if he got the dink. McKinley gave me a thumbs-up.

The rest of the flight was uneventful. We flew by farmers tending their fields and villagers going about their daily routine. They seemed to know not to look up and draw attention from the warbird in the sky as it passed. They probably all knew the Stinger rules of engagement.

I had never been airsick, but I noticed some queasiness on that first mission. As we were returning to base, the queasiness increased.

I thought I was going to upchuck. I tried not to think about the greasy bacon I had for breakfast. I fought to keep it down. I knew I was being watched to determine if I had what it took to be a Stinger. Airsickness on the first mission would imply weakness.

We arrived back at base and as the ship settled to the ground; the queasiness subsided. If fear caused it, I could not detect it. It might have been fear or anxiety on a different level than I had ever experienced. I was never queasy in the air again.

I passed the test. I did not freeze or hesitate when we engaged the enemy.

Sgt. McKinley turned out to be the heart and soul of the enlisted Stingers. I think he did a good job of keeping Stingers on track to accomplish the mission. He appeared to care deeply for the platoon and the reputation the Stingers had built in the Cu Chi area of operations (AO).

The Stinger Crew Chief Duty Station

CHAPTER 10 STINGER MEMORIAL SERVICE

After supper, several of us returned to the hooch. At 1845 hours, all Stinger crew members gathered and walked in a broken formation to the mess hall.

I felt proud to be part of this group of air warriors.

As we entered, there was a pedestal set up at the front of the hall with a flight helmet placed at the center. A chaplain in ceremonial attire was standing nearby, silently waiting.

A few enlisted men and officers from the other flight platoons were also there. We took our seats.

At precisely 1900 hours, the chaplain began the memorial tribute for Specialist Fifth Class Mark Rowland, killed in action. It was a stirring tribute to Rowland, and all fallen comrades of that distant, unpopular war.

When the chaplain closed, all Stingers rose and formed a single line perpendicular to the pedestal. Each Stinger stepped up, one by one, and saluted the fallen Stinger's flight helmet.

Specialist Rowland's body had probably been processed through the in-country Graves Registration facility and was on his way home.

His grieving wife and two young daughters awaited his arrival.

The somber mood subsided as the night wore on.

I thought about how many times they had performed a memorial service like that in Vietnam for fallen American soldiers, airmen, and sailors.

August 24, 1970 - *University of Wisconsin–Madison, Sterling Hall. In protest of the Vietnam War, a bomb was detonated aimed at the U.S. Army Math Research Center on the 2nd, 3rd, and 4th floors of the building. A Ford van packed with explosives missed its target and hit the physics laboratory on the first floor, killing a young researcher and seriously injuring another person.*

August 29, 1970 - *The Chicano Moratorium against the war began in East Los Angeles. A parade through the streets quickly became violent and three days of rioting followed, spreading into South Los Angeles and Wilmington, California.*

CHAPTER 11 FIRST KILL

The next day was another search and kill mission in a different section of the Cu Chi tunnel complex.

The Viet Cong started digging the tunnels by hand with simple tools during the French occupation in the 1940s. They expanded the tunnels during the American occupation. After the war, it was discovered the tunnel complex was 155 miles long.

I took over as Crew Chief that day on Aircraft 523. I got an early ride to the flight line to do a thorough pre-flight inspection. I was not comfortable with McKinley's lead of not inspecting. I am sure the pilots did a good job of pre-flight inspections, but I also was trained for that purpose.

My life, like the other three crewmen, depended on the old bird getting us home. If the pilots wanted to inspect also, that would be fine.

The rest of the crew arrived about forty-five minutes later. The AC was Chief Warrant Officer Second Class (CW-2) Tom Martin and the Co-Pilot was First Lt. James Sanders. SP4 Tony Bremer would be the right side Door Gunner.

Mr. Martin asked me about the ship, "How does the old bird look, Chief."

I replied, "The ship looks good, sir. I see she has a couple of bullet holes in the tail boom that were patched recently. I pulled the inspection plate and double-checked the tail rotor pitch control cables and driveshaft. The bullets missed the critical parts and exited at the top. The main rotor pylon is fine. No excessive play in the push-pull rods. The swashplate and stabilizer bar are good. The particle separator is in good shape. All oil levels are good. No detectable fluid leaks. No bird nests in the turbine exhaust. The tail rotor pitch control rods are OK. I don't see any bullet holes in the rotor blades."

Mr. Martin smiled. "Sounds good Chief, let's load up and go kill some dinks."

I responded, "Yes, Sir."

Mr. Martin added, "Chief, you can call me Tommy."

I replied, "Thank you, sir."

Pre-flight inspection reports were never that formal again. I knew I had to make a good impression on my first mission as Crew Chief.

It became common knowledge among pilots that I was a good crew chief. I guess that is why my ship often got the best aircraft commanders and frequently flew as the lead ship on missions.

The last test was to show I had what it took to kill dinks.

The ammo belts for the M-60 machine guns were stored in ammo cans, which sat on the deck between our legs.

Bremer and I checked the ammo belts for twists or broken links that could jam the belt feed to the weapons.

The first 7.62 mm round is aligned in the M-60 bolt assembly and the top cover is slammed down to lock.

It was interesting to learn the factory ammo belt chute on the M-60 was not used. Instead, a C ration tin can fits the side locks for the ammo chute. The tin can provided more flexibility for the ammo belt feed when the gunner was hanging out of the ship suppressing enemy fire.

The factory metal chute had ridges on the side to keep the ammo belt centered for a stationary ground gunner. In aerial battle, these ridges could jam up the belt feed when the gunner was in extreme positions.

Of course, we removed the tripod assembly and the shoulder butt on the gunship version of the M-60. We always held the stripped-down M-60s on our laps, ready to engage the enemy within seconds.

Ammo belt links fell apart, falling on the deck or blowing out the doorway when the rounds were fired. It was common for the deck to be covered with spent brass cartridges and ammo belt links.

The M-60 effective range was 1,200 yards, with a cyclic rate of fire of 550 rounds per minute. Every fifth round is an orange-red tracer, which is critical for aiming the weapon.

We crossed the base perimeter and switched all weapons to firing mode. I felt no fear. The months of training and psych conditioning kicked in.

I wanted to kill dinks.

We flew along ridgelines that had scattered openings to tunnels. The AC tried to stay just above the tree line at around sixty to eighty knots, so we could surprise the dinks in the open.

In that profile, we were difficult to hit from the ground with small arms fire. The dinks could hear us coming, but a low-level flight over the jungle diffused the sound, masking our position.

A clearing appeared, and a dink opened fire. I saw the AK-47 green tracers streaking towards us milliseconds before I heard crack-crack-crack.

I keyed the intercom and shouted, "Taking fire on the left."

The AC responded, "I got him," as he sharply banked left.

I was about to fire when the main rotor blade arc swept down into my line of fire. I had to wait for a few seconds before I could fire. The AC was already screaming, "Get him, get him, and get him!"

I opened fire.

The dink was running. The AC adjusted the flight path and angle to line me up with the dink. The dink turned and raised his weapon towards us. I was following my tracers and compensating for airspeed to hit my target.

My rounds raked across the dink's body, opening his abdomen, spilling his intestines as he took a few more steps and dropped dead.

We continued the mission.

I had no regrets about killing my first enemy combatant. I had an unfamiliar feeling. It was a mixture of satisfaction of accomplishing the objective of my training and the odd feeling of power to take an enemy's life. Dehumanizing the enemy had removed any misgivings about killing another human being.

We only had about ten minutes left before we needed to turn back to base for fuel.

We came upon a village. All inhabitants scattered. We saw several running into hooches. These primitive structures often concealed openings to the tunnels. I saw a dink with a weapon.

Suddenly, we were taking fire. The AC banked to the left, and I opened fire. The dink with the weapon ducked inside a hooch.

It was open season on hooches.

Both Bremer and I sprayed the hooches with M-60 fire. There was no resistance to the M-60 bullets as they ripped through the grass and bamboo-covered hooch walls. Any occupants, regardless of combatant or non-combatant, had a high likelihood of being killed or wounded.

The AC came around and lined up the ship's rockets with the first hooch. I quickly checked to make sure my foot was not dangling in front of the rocket pod. The AC fired. Something flew up and hit me on the hand. Rockets from both sides swooshed away to the hooch on target. The rockets hit almost simultaneously.

The explosion disintegrated the hooch.

Then there was another immense explosion. The AC sharply pulled the ship up and away to avoid the rising explosion plume and debris. It was a secondary explosion of ammunition or explosives being stored in the hooch.

The AC called the action into the area's ground commander. They organized a ground insertion of 25th Infantry grunts.

We were low on fuel. The 20-minute fuel light had illuminated. The AC turned back to base.

I had a minor cut on my hand. Turned out, metal clips on the back of the rockets would come flying out behind the rockets and enter the cabin area.

It would not be the last time I got hit by rocket clips.

They dispatched another Stinger gunship with a wing ship to cover the infantry insertion.

After we landed and refueled, I started a habit that would stick with me for the rest of my flying days in Vietnam. I patted 523 on her tail boom and mentally thanked her for getting us home.

Stinger Crew Chief View of the Cockpit

CHAPTER 12 LOVERS AND WATER BUFFALO

The constant adrenaline and excitement were still new. It caused me to focus on doing the best job I could. I was living my dream.

It would be weeks before the reality of killing and frequent near-death experiences would weigh on my psyche.

There was no official recognition of my completion of the Stinger acceptance criteria. I just kept getting missions, so that was good enough for me.

I was a real Stinger.

A single ship often conducted search and kill missions in the Cu Chi area without a wing ship. I believe the thinking was, if a ship got shot down, help was not far away.

I did not fly any Landing Zone (LZ) insertion missions in the Cu Chi area. Most missions were search and kill with two missions to support the 25th Infantry ground troops.

My jungle vision improved. I saw more dinks and more opportunities to kill.

My AC that day was Lt. Samuels. He enjoyed killing dinks and pleasing superiors with a high body count.

We were flying over a triple canopy jungle when we streaked past a small clearing. I thought I saw two figures under a tree.

I keyed the mic, "Lt. Samuels, I believe I saw two targets in that small clearing we just passed."

The AC responded and banked in a steep left turn. Our altitude was around 150 feet AGL (Above Ground Level).

The AC slowed the aircraft. It was a tricky balance with a

small clearing. I needed time to acquire the target, yet we did not want to be sitting ducks.

Both figures were still there, they did not run. The AC orbited the clearing and reduced altitude to the top of the canopy. I got a good look when we came around where they were standing. One of the figures was a woman, the other a man. It appeared they were embracing each other.

Then the woman walked away. I did not fire on the woman. She was not running and had no military gear.

The man did not move. He had military gear. I opened fire. My 7.62 rounds peppered his body. He fell to the ground.

In the next orbit, I fired again, finishing the job with hits to the head. The head shattered, spraying brain matter and the red mist of blood over the enemy's twitching body.

We continued the mission.

I often think about that kill. Was the male a deserter meeting with his wife or girlfriend? Had he lost the will to fight? Did he choose to remain stationary and be killed so his loved one could walk away?

I will never know.

The "kill if they run" rule seemed excessive. If a lethal warship is circling above your head, you are probably going to run if you are a combatant or not.

But that was the Stinger way.

I scoured the countryside looking for targets. I made life or death decisions several times during a mission. Are they running, do they have military gear, am I unsure about their combat status, or are they passive, no military gear, and not running?

That was the criteria for life or death.

If they were shooting at us, it required no decision. Return fire.

It sounds strange, but it was a minor relief to get shot at. I did not have to make a life-or-death decision.

The VC used water buffalo like trucks to transport supplies and war material. The farmers used the massive beasts to pull the plows to cultivate their crops.

The farmers willingly or were forced to give their harvest to enemy combatants.

In either case, the animals were precious to the enemy. That made water buffalos military targets.

We were nearing the end of the mission day. We flew across a large parcel of rice paddies. As we crossed the opposite boundary, we spotted a few hooches with a corral holding about twenty water buffalo.

Now, a farmer might have one or two water buffalo, but not twenty.

I keyed the mic, "Lt. Samuels, all of those water buffalo, may support the enemy. If you will orbit, I will take them out."

The AC hesitated for a moment and then banked left. I started firing. We made three orbits. There were no buffalo left standing.

I was getting the feel for this killing business.

That day was my birthday.

I turned nineteen years old.

CHAPTER 13 CU CHI INCOMING

My increasing combat experience was changing me. I was no longer an FNG. I was on the cusp of being an experienced combat veteran.

I was feeling a semblance of immunity from harm. I was getting shot at, yet I was accumulating body count. I was winning.

I was told about incoming artillery from the enemy. It could be short-range mortars or longer-range rockets.

They instructed me on what to do if we had incoming.

There were several bunkers built in the company area. They constructed the protective structures with large metal culverts covered with hundreds of sandbags.

The bunkers would not protect you from a direct hit.

It was common for individual soldiers and command to hire local villagers to perform everyday maintenance and cleaning tasks.

Some female locals kept the officer and enlisted hooches clean. We called them hooch maids. A few hooch maids would double as prostitutes and provide other services.

It was a simple route for enemy infiltration.

Our local workers could walk off distances from the perimeter to prime targets and note a bearing. We caught and dealt with some of these infiltrators.

After months of cutting U.S. soldiers' hair and providing straight razor shaves, we discovered the local barber walking off distances to targets.

We couldn't catch them all.

This valuable intelligence enabled the enemy to accurately target their artillery.

One evening, we heard what sounded like a launch and then a shrill sound coming our way. A 122 mm rocket hit and exploded very near our hooch.

The sound and concussion rattled every cell in my body. All the other guys were up and gone, as my mind was attempting to process.

I heard another launch. I bolted through the doorway and dived into a nearby bunker on top of a couple of guys.

The explosion was on the other side of the company area.

The word "scared" cannot properly describe the reaction to a rocket attack.

There were five or six rockets fired at us that night. A few guys were killed or injured.

My psyche changed. A higher level of subconscious fear formed that would increase with time.

After the war, it was discovered the 116th AHC area at Cu Chi sat on top of the Cu Chi Tunnel complex. They later identified the tunnels as the center of strategic and tactical support for the enemy.

While we slept, the enemy was preparing the next deadly attack right below us.

Limited Protection from Rocket Attacks

CHAPTER 14 AIRCRAFT 523 GETS HIT

There were a couple of rituals observed in the Stingers. They characterized the first as "getting your cherry popped". You lost your virginity when your ship took its first hit after you took over as Crew Chief.

The second ritual was when or if you put a bullet hole in your main rotor blade when firing at the enemy.

The penalty for both was the CE had to buy a case of beer for the platoon.

I got my cherry popped after a few weeks with the Stingers.

A Stinger gunship was hit the day before, west of Cu Chi. The crew got the ship home. My ship, 523, and a wing ship were tasked with going back out to the same area and seek revenge.

My right-side gunner would be Bremer. Our AC that day was none other than "sure shot" CW2 Lou Baldassaro. He insisted we call him Lou.

I had met Lou a few days prior at the Stinger hooch. We were almost the same age, except Lou looked like he was fifteen years old. He had a baby face that would make any mother proud.

Lou's appearance was deceiving. He had become the most lethal and accomplished pilot of all the Stinger ACs.

It was an honor for Lou to agree to fly with a CE. So, I considered the event as a milestone in my Stinger career.

I felt like I had "made it" when Lou took my word the ship was ready to fly. Neither he nor the co-pilot did their own inspection.

We cleared the base perimeter and activated all weapon systems.

The top Stinger AC was in command. We were ready to kill

some dinks.

We were flying along a river at about 100 feet AGL. I smelled a terrible odor. It was worse than the usual smell of rotting animal flesh. It was a distinct smell.

The smell was getting worse. Lou keyed the mic and said, "Man, that stink is intense. We need to gain some altitude."

As 523 climbed to 200 feet AGL, we could still smell the rot.

We spotted an enemy corpse floating in the river. The dink had bloated out to three times its normal size. Lou slowed down to check out the corpse. While directly above the inflated body, we were hit with the concentrated smell of human death from 200 feet away.

If you ever smell rotting human flesh, you will never forget the unique odor.

We arrived in the target area and started our tandem patrol. Our wing ship trailed to our right.

After years of Stinger domination in the Cu Chi AO, the sight of two Stinger gunships bearing down must have put the fear of certain death in the hearts of the most battle-hardened VC.

Lou had developed a sixth sense in locating VC positions. Within five minutes of being on station, the sound of a hammer pounding on our tail boom rang out. Then, we heard crack-crack-crack!!

Multiple dinks with AK-47s were shooting at us. We were hit.

Just as I screamed, "Taking fire" on the intercom, Lou had entered a steep left bank. I was almost looking straight down horizontal with the ground as we entered the high-angle turn.

The main rotor arc completely obscured my higher angle field of fire.

We were still taking fire. I thought, "Fuck it" and returned fire through the rotor blade sweep. The odds were in my favor. I would not hit a rotor blade.

I spotted three dinks in a ravine firing at us. I adjusted my firing path and swept their position. At least one dink head exploded into fragments.

Lou lined up the ship to fire rockets into another enemy position. Both the right-side gunner and I were putting down suppressing fire.

I caught a glance of our wing ship through Bremer's doorway. They had found their own nest of VC and were in full engagement.

Lou fired rockets at several dinks, who were shooting at us from behind some trees. Right on target. The 17-pound warheads exploded, shattering the trees.

Two partial VC bodies spiraled away in opposite directions.

Lou and I both saw a green tracer from a dink AK-47 coming straight for us. We both watched the tracer getting closer and closer. In real-time, there was nothing we could do.

The tracer seemed to slow down. Lou and I were both mesmerized by the bullet, as it appeared to be right on course to hit one of us. Physics was on our side that day. The bullet ran out of energy before it could hit us. We both watched the bullet drop away.

Lou turned to me with a big, toothy grin. I just shook my head.

The Stinger tandem team finished the day with eight dinks killed by rockets and five dinks killed by gunners.

Lou's body count soared to 71. That was a good day.

We got back to base and landed. I inspected 523's battle damage and took notes. There were no enemy or self-induced CE bullet holes in the rotor blades.

I patted 523 on her tail boom and silently thanked her for getting us home.

Bremer and I had flown together several times, and he always snickered when he saw me pat old 523 on the tail boom at the end of a mission.

Today, he couldn't help himself. He complained in his southern drawl, "Allen, ya treat the goddamn helicopter like it was your girlfriend or somethin'. It's fuckin' weird."

I smiled and replied, "She is more than my girlfriend. She takes us into battle, gets shot up, and always gets us home."

I reported the required repairs to the maintenance tech in charge. The tech assured me they would have the ship ready to go by morning.

I had one more task to complete. I caught a ride to the PX (Post Exchange). I purchased a case of warm beer. We were in a war. No refrigerated beer at the PX. I returned to the Stinger AO.

I carried the case of beer on my shoulder and entered the Stinger enlisted hooch. I announced I had relinquished my virginity and passed cans of beer all around. Everyone cheered and guzzled away. It was a splendid day.

Stinger 523

CHAPTER 15 COMBAT MENTALITY

A life in combat means nothing.

Life as a civilian means everything.

It is impossible to comprehend the complete reality of a combat soldier with a typical civilian psyche. One or more daily near-death experiences were routine in gunship platoons.

If that reality could be transposed to an average civilian psyche, the gravity would be overwhelming.

The constant lethality and required courage are unfathomable for most civilians.

Front-line combat soldiers adapt a unique perspective to endure. We just accepted the high probability we would not survive the war.

The prevailing attitude was, "don't mean nuthin".

Like most soldiers, my girlfriend and I exchanged voluminous letters filled with yearning to be together again.

My psyche gradually changed to that of a frequent killer. Enemy bullets continued to hit my aircraft, sometimes inches from my position. The letters I wrote became hollow.

I found it difficult to convey my experiences. I soon stopped writing about the war and placated my girlfriend with words she wanted to hear. Her letters were a much-needed respite from the war, so I wanted to keep communicating.

The young man she fell in love with no longer existed.

The warrior's mentality of killing the enemy was synonymous with sport. The adrenaline rush of preparing for a mission, knowing you would be shot at, possibly wounded, or killed with the reward of adding more body count to your record, was intoxicating.

High enemy body count and surviving were comparable to winning the league championship.

When I transitioned back to civilian life, it surprised me I no longer had the desire to hunt animals. It was a sport I enjoyed as a kid. As a combat veteran, I no longer saw any point in killing an adversary that was not capable of shooting back.

The allure of sport was gone.

The incoming rocket attacks added another dimension to our tortured mental state. The massive violence of rocket explosions unleashed unquantifiable fear. The result was soldiers became zombies.

Unfeeling, uncaring, numb, stone-cold killers. Losing the sense of humanity helped.

Unit camaraderie is an essential component for combat soldiers. The one for all, all for one premise, helped a soldier to endure and excel.

In late 1970 and early 1971, the Vietnam War was ending. Our Nation was attempting to extricate the warriors without a clear-cut victory.

Thousands of American soldiers sacrificed their lives and survivors lost future well-being for a cause without purpose.

The ensuing quagmire for the soldiers still fighting, killing, and enduring near-death events, while the troop strength decreased, created a noxious environment.

Unit camaraderie suffered.

I removed the small Bible from my flight suit pocket and stowed it away in my footlocker. I felt like I was at the point where I was beyond providential protection.

When combat soldiers leave the war and become civilians again, it is common for them to become emotional when they talk about their war experiences. It is embarrassing, so most combat guys just don't talk about it.

I felt a little better about my emotions when I saw the video of the great NFL Pittsburgh Steelers running back and Vietnam vet, Rocky Bleier. He became emotional when he revisited Vietnam after the war.

A man can't get much tougher than Rocky Bleier. If his war experience was still emotional for him, I guess I won't beat myself up for shedding a few tears.

I have often thought about how we endured day after day in combat, yet were not outwardly emotional about the horrors. I realize now, visible emotional control is subject to your current mental state.

When you are in the war and are often in life or death situations, you are in combat mode. You are numb to civil niceties. You have little or no sense of humanity.

You are a warrior, and your job is to kill people.

It is a dark psychological place to be in, but it is necessary so you can do your job.

After a few years back with family and friends in the "world", humanity seeps back into your psyche. The re-emergence of humanity sharply conflicts with your combat experience. Your reconditioned state of mind is appalled and terrified of those war memories. The fear, anger, and ferocity attached to war memories are sometimes just too much for a recovering psyche.

CHAPTER 16 SCRAMBLE TO SUPPORT the 25th Infantry

It had rained almost the entire day. Operations had scrubbed our usual search and kill mission. We were settling in for the evening when the field phone rang in the Stinger hooch.

Phil Parsons, our crew chief from Texas, answered the phone. His demeanor changed.

We heard him say, "Holy shit. I got it. How much time do we have?"

Operations replied.

Parsons slammed down the phone and yelled, "523 and 772 are scrambled to save a platoon of grunts about fifteen miles away. The situation is critical. Operations said it may already be too late."

The two crews of gunners burst out of the hooch into a dead run to the waiting trucks. The pilots arrived moments later. We all piled into the two trucks and launched.

Lou Baldassaro would be the AC for my ship, 523

As we were thrown around in the back of the truck, swerving around potholes in the dirt road at a high rate of speed, Lou gave us some details.

"A 25th Division infantry platoon is nearly surrounded. An overwhelming enemy force is threatening to overrun the grunt position."

American soldiers were dying. With the lives of the infantry guys in grave danger, there would be no preflight inspections.

The M-60s were always left in standby readiness.

Bremer and I jumped off the truck before it had stopped as it neared 523. The pilots were right behind us. The other truck carrying the wing ship crew arrived at their ship simultaneously.

There was still a light rain. We had some daylight left. With no delay, everyone did their job. I unhooked the main rotor tie-down and checked around the ship for obstructions.

Bremer quickly checked the guns and rockets. The pilots scanned the essentials of the preflight checklist.

The AC fired up the ship, the gunner and I cleared and we lifted off. The wing ship was right behind us.

As we neared the infantry platoon's position, they popped a smoke grenade. Our AC identified the color to ensure the enemy had not intercepted the radio transmission and popped their own smoke.

The grunt lieutenant confirmed and gave us vectors to three enemy locations.

He estimated a force of 50 to 60 dinks. The enemy was close to the Americans. It would be tight to deliver lethal fire on the enemy and not hit our troops.

We started taking fire. The enemy tracers were bright in the early evening sky as they flew past our ship.

The AC lined up the ship on the first enemy position to fire rockets. Bremer and I were hanging outside the ship with both M-60s blazing as enemy tracers streaked by.

We completed our rocket run and watched the explosions rip the jungle apart. Enemy fire had slowed from that position. The wing ship rolled in from behind us and repeated the attack.

We started a left-hand orbit to intersect the next enemy

position.

The ground tracers were intense. We were taking hits.

A bullet broke through the right side pilot windscreen and streaked a few inches above the top of the co-pilot's flight helmet. The bullet went through the top of the cabin about a foot in front of Bremer. Bremer turned and looked at me with a half-smile, then continued spraying the enemy with his M-60.

After the wing ship completed its rocket run, the American infantry went into full counter-attack as they overran the first enemy position we had just assaulted.

The AC lined up to fire rockets at the second enemy position. Bremer and I shifted our field of fire forward.

The M-60 barrels were hot and smoking.

An RPG (rocket-propelled grenade) launched from the ground towards us. We watched as the missile flew wide of us.

The AC fired the rockets. We sprayed the area with M-60 fire. We continued left-hand orbits around the enemy positions.

Lou was communicating with the infantry lieutenant to coordinate the counter-attack.

The remaining enemy force withdrew.

We searched the area. Bremer and I laid down a steady stream of machine-gun fire to draw the enemy out.

No luck. The surviving enemy had escaped and evaded.

That is the reason the enemy liked to strike in the early evening. They knew the avenging gunships would come to protect their own. As darkness fell, the enemy could escape into the night.

Grunts were dying from wounds. Dustoff (medical evacuation Huey) had been scrambled from Cu Chi and was on the way.

Our wing ship was losing hydraulic fluid from bullet hits and had to break off.

We stayed on station to cover the Dustoff landing and departure.

Two infantry guys were KIA and four were wounded. They were part of the eighteen-man platoon.

The infantry lieutenant radioed and thanked us for our intervention. He offered to buy drinks at the officer's club for the pilots if they ever met.

My ship, 523, took three hits. The wing ship took five hits.

No aircrew was injured.

Of course, 523 got her appreciative pat on the tail boom for getting us home.

I later heard there were fifteen dinks killed by the gunships and more killed by the grunts.

M-60 machine gun ready to go

CHAPTER 17 REASSIGNMENT TO I CORPS

Life for the 116th AHC in Cu Chi was as comfortable as you could be in a war. We had a swimming pool, bar, volleyball court, and other amenities.

Everyone knew who we were and what our history was.

They respected us within the 25th Infantry Division. Many grunts went home alive thanks to the Stingers, Wasps, and Yellow Jackets.

The thing about being a proficient assault helicopter company, the Army would send us where they needed us the most, with brief notice.

Counter-Offensive Phase VII started on July 1, 1970. The overall objective was to reinforce I Corps along the western and northern border of South Vietnam. The NVA was moving its supplies from their homeland into South Vietnam along the Ho Chi Minh Trail, just across the South Vietnam western border in Laos.

The mission was to cut that vital supply line and build up the force in I Corps to protect the country at its northern border from an NVA mass invasion.

U.S. Command was maneuvering its assets to assist the ARVNs (Army of the Republic of Vietnam) to take over the war so Americans could extricate themselves from the quagmire. It was called "Vietnamization".

The U.S. Infantry division in I Corps was the 23rd Infantry called Americal. The same Americal of the My Lai infamy.

The news of the 1968 My Lai massacre hit global headlines in late 1969. Army high command nailed the Americal commanders to the cross before they were reassigned to some place worse than Vietnam or forced to retire early. They had most likely warned the new commanders that attracting such intense negative publicity would not be tolerated.

Army Command reassigned the 116th AHC to Chu Lai, home of Americal headquarters. Our company area was a primal beachfront "resort" right on the South China Sea. The ocean waves were stupendous. Body surfing was world-class.

The downside was, we had a huge fuel tank farm right next to us that stored millions of gallons of flammable JP4 jet fuel.

To resupply all those gigantic tanks, ocean-going tankers would offload the fuel through a pipeline about a quarter of a mile offshore from our company area.

The entire time we were at Chu Lai, no one ever bothered to fix leaks in the pipeline. Judging by the amount of fuel that was in the water, there were probably hundreds of thousands of gallons of JP4 being lost at U.S. taxpayer expense.

So, like California surfers who waited for the call, "surf's up", our group of haggard-looking body surfers waited for the days when there were no tankers.

We also needed a storm to come through, creating high waves and cleansing our part of the ocean of JP4.

The tank farm had another rather negative side effect. The beach line jutted out to sea on the other side of the tank farm. The VC would haul their rockets to the outcrop and launch them at night to blow up whatever they could hit.

They would often overshoot the tank farm and hit our company area.

I grew to fear incoming rockets far more than being shot at in my gunship.

To round out the threat assessment, on moonless nights, the

VC would take to the ocean in homemade boats that were just big enough to carry one dink and a satchel explosive. We called these guys "sappers".

The sappers would attempt to plant the explosives on the tanks or one of our company area hooches.

There were stories of sappers sneaking into U.S. hooches and slitting the throats of sleeping soldiers.

Putting all of that aside, the 116th was coming into a new division, where we had no relations or reputation and no amenities except the JP4 contaminated ocean.

Entrance to 116th AHC company area

Shower and shaving station

116th AHC Beach - South China Sea

The southern beach guard tower in the distance

Stinger Enlisted Hooch

116th AHC makeshift urinal

116th AHC Company Area

1971 Monsoon

CHAPTER 18 KHAM DUC

The 116th was getting squared away in our new company area. Just a few days after arrival, they circulated the announcement there would be a mandatory company formation for enlisted personnel the next morning.

This would be the first of only two mandatory company formations during my year with the unit.

At 0800 the next morning, all the enlisted flight platoons, maintenance, and other support personnel lined up in rank and file.

The new First Sergeant addressed the formation. "Men, it is my duty to inform you we will conduct a joint mission of several units tomorrow. It is highly classified and extremely dangerous. We expect to lose one-third of our flight personnel."

The First Sergeant became emotional. He finished his announcement and advised the mission would begin before dawn.

I would be lying if I said I was not affected. I was becoming accustomed to being shot at and a little calmer about rocket attacks.

But this proclamation scared the hell out of me. It was time to suck it up.

The next day, in the pre-dawn darkness, the Stingers loaded up on the trucks and headed for the flight line.

As I relive that memory, I remember feeling the gravity of that mission as our truck sped along the dirt road.

The First Sergeant's claim that one-third of us would die that day still echoed in my mind.

We were like the aircrews in World War II, riding their jeeps to the waiting bombers in the early morning for precision daylight bombing missions over Germany. They knew they would lose many of their aircraft and crews.

Flying in combat aircraft is a double threat. Bullets or shrapnel could kill you. Even if you were unscathed, you could die in a burning, out-of-control aircraft falling from the sky. It is a feeling only combat aircrews of any war can understand.

Our truck pulled up to 523. I had done my pre-flight inspection the prior evening. The pilots arrived after their last briefing. The AC said we were going to a place called Kham Duc near the Laotian border.

I was unaware of Kham Duc's history. I later learned, in the 1968 Tet offensive, Kham Duc was the only clear victory for the North Vietnamese armed forces and the VC.

The enemy's victory was short-lived. The next day, our B-52 bombers turned the area into a moonscape.

The NVA regained control. In August 1970, Americal decided to take it back. It was a necessary strategic move to prepare for the upcoming major assault in Laos.

They tasked assault helicopter platoons from several units

The new guys, the 116th AHC Hornets, were expendable, so they ordered us to lead the air assault. If we survived, Americal would commit all their units.

There was no current intelligence to estimate enemy strength and positions. Americal opted not to gather current intel by sending in a recon team for fear it would alert the enemy of the imminent assault. The extent of enemy resistance was unknown. The 116th would be the sacrificial reconnaissance force.

Fifteen Yellow Jacket, Wasp slicks, and six Stinger gunships got airborne, formed a tight formation, and proceeded to Kham

Duc.

It was an impressive, heart-stirring sight. Twenty-one Hornet assault helicopters led the charge, backed by several Americal assault helicopter units. It filled the early morning sky with warbirds headed into harm's way.

The terrain was mountainous, with peaks reaching thousands of feet. Below was a triple canopy jungle. Emergency landings were not a viable option.

As the armada approached the Kham Duc airstrip, the Stingers surged ahead and began rocket runs. The gunners laid down suppressive machine gunfire on both sides of the airstrip. The intent was to detonate mines or booby traps and slaughter any defending NVA.

The Hornet slicks came in fast, then pulled up at the last minute to allow the Americal grunts to jump to the ground and take defensive positions.

There was no enemy resistance. The LZ was cold.

We completed several more insertions that day. Americal grunts searched the area. Few NVA were found.

It was unlikely the NVA would not have any units in such a strategic location. Intelligence leaks seemed to be a problem everywhere.

Although we had no enemy contact that day, I had become more superstitious about giving 523 the appreciative pat on the tail boom. She could have quit on us and forced a crash landing in the middle of enemy badlands.

She deserved a pat on the boom.

CHAPTER 19 THE STINGERS AND MY LAI

After the Kham Duc fiasco, we realized our new Division may be more of a challenge than we had expected.

The Hornets still wore the coveted First Aviation Brigade patch, because no orders were issued attaching our unit to Americal. So, we were just the red-haired stepchild on loan to Americal.

My ship had to go in for a 100-hour inspection, so I got in some beach time. When 523 came out of maintenance, I expected to get a mission. It was a shock to learn all 116th AHC flight platoons were grounded for several days.

We were wondering what the hell was going on. We later learned the grounding of all Hornet forces was punitive.

The enemy in I Corps was accustomed to a response delay from Americal units when contact with the enemy was made. Americal command allegedly required infantry and aviation units to get permission from higher command to engage the enemy. A knee-jerk reaction from the My Lai massacre.

That was not the Stinger way, and we received no specific orders.

The My Lai massacre led by Lt. Calley and his Americal infantry unit was a tragedy. Not even the Stingers would kill women and children who had no weapons or military gear. But you have to look at Calley's mindset. The area was a known Viet Cong stronghold. Calley, like all combat soldiers in Vietnam, was under implied standing orders to accumulate body count.

The theater-wide strategy was to win the war by attrition. Kill as many enemies as possible until the adversary could no longer sustain its fighting force.

Calley probably believed all the villagers in My Lai were involved with the Viet Cong. He believed he was doing his job. He crossed the line.

After the massacre, political pressure forced Americal to avoid My Lai. Eventually, the enemy realized My Lai was "protected" and would be a haven for VC operations.

Americal sent our Wasp slicks to an area near My Lai accompanied by Stinger gunships. The mission was to extract a platoon of Americal grunts.

There were no unusual limitations attached to the mission.

As the slicks landed in the LZ for the pickup, they began receiving incoming mortars. Instead of focusing on the source of the mortar fire, the grunts opened fire on the entire perimeter. So, the mortars kept raining down and exploding.

The Wasps could see the enemy firing position in the nearby village. The Wasps directed the Stingers to the position to take out the mortars.

The Stingers rolled in on the village, firing rockets and door gunner machine gunfire. The threat was extinguished.

The Wasps lifted off for the return to base. After they delivered the grunts, the Wasps were instructed to return to the Hornet base and shut down. It was early in the day, with only one sortie. That was very unusual.

Our Commanding Officer and Executive Officer (XO) were ordered to report to the Americal Commanding General.

The village the Stingers had attacked to take out the mortar position was My Lai.

The news of the Stinger attack on My Lai was explosive and far-reaching. Americal reacted as if the Stingers had gone "off-book" and conducted an illegal operation. We annihilated any hope of forming a good relationship with Americal.

Allen getting a few days off

They ordered the 116th Hornet Command to undergo strict indoctrination of the Americal rules of engagement.

Orders were cut to formally transfer the 116th AHC from First Aviation Brigade to Americal. They ordered Hornet personnel to immediately remove the endeared First Aviation Brigade patch from our uniforms and replace it with the Americal patch. The transfer gave Americal full control of the 116th.

Our Company Commander was relieved of his command nine days later.

The following week, a new black crew chief was assigned to the Stingers. His name and rank was SP5 Alvin Jones. We had no minorities in the Stingers, so it was different. Most Stingers had no issue with Jones. I certainly didn't. He turned out to be a friendly guy and a great crew chief.

My usual door gunner, Bremer, however, had a deep-seated resentment for blacks. I guess that was just part of his upbringing in the Deep South. Bremer did not hide his prejudice.

SP5 Jones was unfazed. He took it in stride.

October 7, 1970-*In a nationally televised address, President Nixon proposed a five-point peace plan for Indochina. The Plan included a "ceasefire in place" and the negotiated withdrawal of U.S. troops from Vietnam.*

CHAPTER 20 174TH AHC SHARKS INITIATION

The 174th AHC became part of Americal a few years earlier. They were based south of us at Duc Pho. We had heard of the 174th gunship platoon, called the Sharks. They had an impressive combat record and a fearsome reputation.

My ship, 523, was tasked to fly to Duc Pho and join the Sharks as a wingman on a standby mission.

Our AC was CW2 Donald Smith "Smitty" and co-pilot Lt. George Peters. Bremer was my door gunner. We arrived at the Duc Pho base in the late afternoon. A 174th truck was on hand to ferry us to the company area.

The officers went off to do their thing. Bremer and I searched for the Shark's enlisted hooch. With a little help, we found it.

The Sharks warmly greeted us. The exchange of mutual respect was immediate. The Sharks were our brothers.

Even though we were on standby and could be called out for a mission at any time, the Sharks produced booze in copious amounts.

It didn't take long for the booze to loosen the trash talk.

Sensing I might get an unbiased answer, I had to ask, "How did the Sharks handle Americal's restrictive rules of engagement?"

The response was, "radio failure". If a Shark took fire, they returned fire. No questions or requests to return fire were needed. If they were questioned by command, they claimed radio failure.

I asked about dinks who were not shooting at you. The Sharks roundly responded, "All dinks are enemy or support the enemy. Kill them all and let God sort them out."

I attributed that response to the booze.

There appeared to be little or no difference between the Stingers and the Sharks regarding rules of engagement.

The only difference was, right off the bat, we had unknowingly prodded the Americal festering wound aka My Lai.

As the sobriety levels deteriorated, the Sharks talked about how the locals reacted to a Shark gunship passing by. The women would strip naked.

Like the locals in the former Stinger AO around Cu Chi, the Duc Pho area locals also knew the gunship rules of engagement. So the native women in the Shark AO went a step further to show they had no concealed weapons or military gear.

Bremer and I were impressed. The Sharks rocked!

As the collective consciousness slipped into a drunken haze, the Sharks decided the Stinger visitors had to be initiated as Sharks.

A Shark produced a human skull. The skull had two small-caliber bullet holes on top of the skull. The skull belonged to a VC sapper that was killed in the perimeter wire about a year earlier.

The Sharks had acquired the skull when it was still fresh. They removed the brain and associated tissue to re-design the skull into a drinking vessel. They would pour booze into the base of the upside-down skull. The inductee would drink the booze pouring out of the two bullet holes.

Fortunately for Bremer and me, the skull had long since dried out and the maggots were gone.

We both dutifully raised the skull above us. As booze was being poured in, we guzzled all we could catch streaming from the bullet holes.

I awoke the next morning with a massive hangover. The gods of war were kind. I could partially recover that morning before we caught our only mission in the afternoon.

After we did our pre-flights and armament checks, the Shark / Stinger gunship team departed to the west.

Shortly, we arrived at a flat area strewn with over fifty bodies.

Enemy and friendly KIAs were near each other. The fight had been ferocious. There were signs of hand-to-hand combat.

The graves registration guys were still recovering the dead Americans.

The Sharks said they were involved in this battle two days ago. I never got all the details.

It was the highest number of war dead in one battle I ever witnessed.

We encountered no targets on that mission. We were released and flew back to Chu Lai.

I never hooked up with the Sharks again.

I know the skull initiation sounds satanic and grotesque to most readers. I ask the context be taken into consideration.

The men involved in the gathering that night had collectively accumulated hundreds of enemy kills. All had dodged enemy bullets and incoming rockets countless times.

After combat soldiers have been exposed to that level of conflict for prolonged periods, vestiges of civility and humanity are completely extinguished.

There was a popular saying among highly experienced combat soldiers.

"Though I walk through the valley in the shadow of death, I shall fear no evil. Because I am the evilest motherfucker in the valley."

This was a necessary mindset that improved the chances of surviving Vietnam as a frontline combat soldier.

World War II combat troops in the South Pacific, on average, saw forty days of combat in four years.

Vietnam's combat troops saw over two hundred days of combat in twelve months.

One out of every 10 Americans who served in Vietnam was a casualty (wounded or killed.)

174th AHC Shark Gunship

CHAPTER 21 TAM KY INSERTION

We got orders to report to the flight line at 0600 hours the next morning. We were told that my ship and a Stinger wingman would cover an insertion of Yellow Jacket slicks carrying Americal troops. The landing zone (LZ) was near Tam Ky, about fifteen miles northwest of Chu Lai.

As a special treat, the mess hall would open early just for the Stinger and Yellow Jacket flight crews, so we could get a quick breakfast.

That raised eyebrows. We were becoming accustomed to Americal putting the screws to us. This sudden gesture of hospitality was suspicious.

At 0515 the next morning, McKinley was kicking mine and Bremer's beds to cheerfully awaken us. We were up, dressed, and on the truck at 0530.

After a speedy breakfast, we arrived at the flight line at 0600.

I did a quick aircraft inspection and Bremer checked the armament.

Our AC, CW2 Tommy Martin, called all pilots and crews of both gunships together for the briefing.

The LZ south of Tam Ky was expected to be hot. They had sighted the NVA in the area with heavy machine guns and RPGs.

The Americal troops' mission was to assault the NVA in the area around the LZ. They would then set up an encampment to conduct operations in the general area to destroy the NVA build-up.

The early morning operation was designed to catch the NVA before they realized what was happening and could get organized.

The LZ had never been used before, but low-level recon flights showed it was suitable for the insertion.

The Stingers would go in first and prep the area around the LZ with M-60 and rocket fire.

At 0700, the Stingers lifted off and headed northwest. The six Yellow Jacket slicks, with their complement of Americal grunts, lifted off about ten minutes later.

As we approached the LZ, the AC keyed the mic and told Bremer, and I to get ready and keep our eyes open.

To me, that was unusual because Tommy knew Bremer and I were always on full alert. It was a sign that Tommy had higher than normal concerns about this operation.

As we entered the first orbit around the LZ, all was quiet. Too quiet.

The wing ship reported movement as 523 passed.

Bremer and I leaned out of the ship, standing on the rocket pods, looking for the enemy. I saw uniformed dinks moving a machine gun into position. I keyed the mic, "Tommy, we have dinks."

The world exploded. Heavy enemy fire opened up on both gunships. The surrounding air was filled with incoming tracers. The gunners on both gunships opened fire.

Tommy nosed 523 over and fired rockets. The wing ship fired their rockets under us. We abandoned the orbit and flew straight ahead to get out of range and regroup.

Miraculously, neither ship had been hit. Tommy radioed the Yellow Jacket commander and reported the heavy resistance. The Yellow Jackets delayed the insertion and started a holding pattern a few miles away. The Stingers would have to take out the enemy positions so the insertion could proceed.

Tommy keyed the mic for the whole crew. "OK, guys, get ready. We are going back in."

The gunships did a 180-degree turn and assumed attack profile. Tommy aimed and fired the rockets. Bremer and I moved halfway out of the ship and started firing forward. We saw several clusters of dinks.

I hollered at Bremer, "Drop grenades as we pass and I will keep firing". Bremer acknowledged and dragged the ammo can full of grenades over to his side. As Tommy and the wing ship were firing rockets, Bremer dropped grenades. I continued firing my M-60 at enemy positions.

The wing ship gunners caught on and started dropping their grenades while firing their M-60s.

We started taking fire from the other side of the LZ. The NVA were ready for us. This was a coordinated setup.

As we completed the run on the left side of the LZ, our wing ship broke off and began a run on the right side of the LZ.

The jungle tree line was about 50 feet from the LZ on both sides. The dinks were using the jungle as cover. The disadvantage with that strategy was the dinks could not get a clear shot at us as we flew by on our rocket and strafing runs. The dinks had to come out into the open to shoot at us as we approached and passed by.

The ACs of both ships laid down rocket fire along the edge of the jungle to kill the dinks in hiding. On the next run, I grabbed the M-79 grenade launcher and fired several aimed grenades.

Both Stinger warships continuously strafed the concentrated enemy positions, dispensing death in every way possible.

Dead and dying NVA lie scattered on both sides of the LZ. Bremer ran out of frag grenades, so he dropped white phosphorous grenades (called Willy Pete). The Willy

Pete grenades exploded and scattered chunks of burning phosphorous that could not be extinguished except by submerging in water to deprive of air. The Willy Pete burns at 5,000 degrees Fahrenheit.

We saw NVA troops running around with Willy Pete burns that could not be extinguished. It was a gruesome death.

Tommy called the Yellow Jacket commander and declared the LZ was cleared well enough for the insertion to proceed. The Americal troops could mop up the operation on the ground.

As the Stingers orbited the landing, I saw the disembarking grunts give the gunships a "thumbs up".

Without gunship intervention, the insertion would have been a slaughter of young Americal soldiers.

Both Stinger ships took several hits. We made it back to base. After we landed and shut down, I performed a thorough inspection of 523 to evaluate the damage.

There were five bullet holes in the tail boom. One bullet had hit the tail rotor drive shaft.

I traced bullet trajectories from the bottom of the ship under the engine compartment. I found damage to the turbine engine near the fuel control module.

Two bullets hit the cabin area. One near Bremer and the other in front of me near the AC seat.

I made notes of the damage to the maintenance guys.

523 got a well-deserved pat on the boom that night.

Allen relaxing after a tough day in Tam Ky

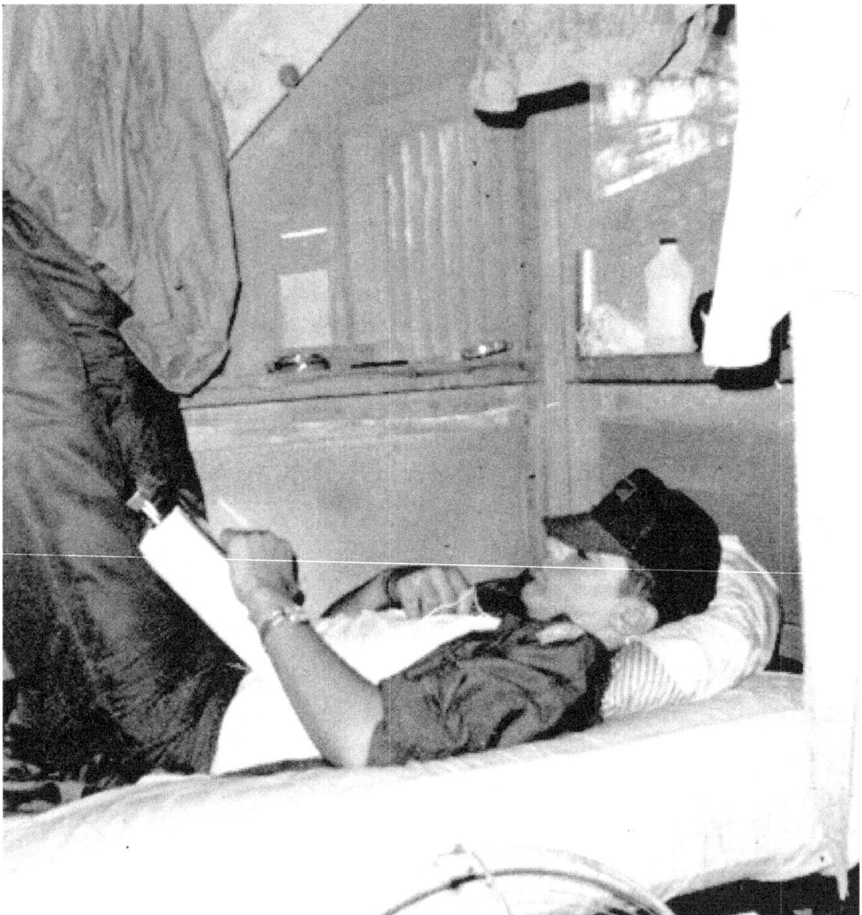

CHAPTER 22 CRASH AND BURN - AIRCRAFT 523

We were walking back to the hooch after having a forgettable supper in the mess hall. One guy started talking about the steak dinner served to Vietnam veterans when they got back to the States.

I had not heard that one before. I was far from being short. So I tried not to think about a sizzling, tender, delicious steak, medium well, covered in mushrooms with a few dashes of A-1 sauce. Perhaps a nice baked potato and hot yeast rolls smothered in melted butter.

OK, that's enough. The best we could get was a dried-out hamburger with the consistency of cardboard served at the PX.

Everyone was settling in for the evening. The field phone rang. McKinley answered. His voice became urgent as he responded, "Right, I got it. The crew chief and gunner are right here. They are on the way."

He slammed down the phone and yelled, "Allen, Bremer, you have a scramble mission. Run over to the officer's hooch and join Smitty and Lt. Cooper."

The pilots busted out the hooch door just as we arrived. Smitty would be the AC. He blurted, "We have a truck waiting at the orderly room. I'll give you the details on the way. Let's go."

We piled on the truck and took off at high speed on the rough dirt road.

Above the roar and frequent potholes, Smitty filled us in. "We have to provide cover for a rescue effort of a downed slick. They have lost radio communication with the ship. We understand four crewmen are aboard, along with a few infantry guys. If there were any survivors after the crash, their time was limited.

A heavy concentration of enemy troops was spotted nearby."

It would be a race to get there before the enemy.

We would have to engage the enemy while a rescue slick attempted to retrieve survivors and bodies. No other Stinger gunship was available to be a wingman. Americal assigned an AH-1C Cobra gunship from another unit as our wingman.

It was ironic since the Cobra was much better armed, with more armor protection, and much faster than our old UH-1C Huey. The AH-1C was a dedicated attack helicopter. It was relatively new with the latest technology. The only similarity was both ships had the same rotor system to enhance maneuverability.

The truck screeched to a halt alongside 523. The crew jumped out of the truck. The pilots climbed into their armored aircraft seats. They immediately began the start sequence while simultaneously strapping in. I removed the main rotor tie-down. Bremer rushed to check the armament. We cleared the area for obstacles. The main rotor came up to speed. We were off.

The two-man Cobra joined us on the way to the crash site.

Upon arrival at the crash, the Stinger and Cobra established a left-hand orbit around the badly damaged slick. The Cobra maintained a higher altitude. If the Stinger took fire, the Cobra would roll in with all of their lethal weaponry, besides the Stinger's killing capabilities.

The rescue slick landed nearby. A recovery team made their way to the crash site. After the team inspected the crash, they radioed their report.

The enemy had reached the crash site before us. It appeared there were survivors from the crash. The enemy had overwhelmed the defensive fire from the survivors. They mutilated the dead or dying defenders. Some guys had their eyes cut out. Two guys had their genitals cut off and stuck in their mouths. Other soldiers had their ears cut off.

The enemy was spotted returning to the crash site.

Command ordered the rescue team to abort and return to their ship. They safely lifted off and departed the area. They also ordered the gunships to abort and not engage the enemy.

We were angry.

After My Lai, Americal would usually not allow gunships to engage at night.

We still had some daylight left.

Americal refused to allow us to engage the enemy. They were afraid of killing non-combatants, like My Lai.

The enemy got a free pass.

Stinger 523 and the Cobra broke off and started back to base. Everyone was silent, thinking about our dead and mutilated fellow soldiers. Their bodies may never be recovered and returned home.

Smitty told the Cobra to go ahead since we were much slower.

We encountered a light rain that soaked my body as I positioned myself halfway out of the ship with M-60 armed and ready.

We cruised along at about one hundred feet AGL. I watched the jungle and rice paddies pass by as I sat in somber solitude.

I heard the co-pilot over the intercom announce our 20-minute fuel light had illuminated.

That meant we had about 30 gallons of fuel left in a tank that held 235 gallons. So, we had a lot more highly combustible fuel vapor in the tank than fuel.

I leaned back inside the ship to look at the instrument panel between the pilots. I wanted to see the torque meter to check our load level.

523 had been lagging, and I wanted to make sure it had not gotten worse.

It shocked me to see the torque meter registering forty pounds, which was the highest load level I had ever seen ole' 523 produce.

I keyed the intercom, "Hey Smitty, the old bird is pulling forty pounds."

The pilots looked at each other and smiled.

She was giving all that she had.

The next moment, everything exploded. Heavy machine-gun tracers burst through the pilot's center radio console and tracked through the door gunner's cabin, piercing the fuel cell behind us.

Another enemy bullet hit the bottom of the ship and broke through the deck about six inches from my right foot. Shrapnel sliced into my leg and hand.

The Aircraft Commander started firing the rockets. The door gunner and I returned fire.

As I leaned out of the ship, I was hit by flames.

I screamed on the intercom, "Smitty, we are on fire. We are going to blow. We have to land!!"

Smitty kept flying to put distance between us and the enemy. Our capture would have resulted in certain gruesome deaths.

Smitty transmitted a Mayday call out to the Cobra and the rescue slick. Both ships could arrive back at our position within fifteen minutes.

I could move my injured leg and hand. None of the wounds seemed to be serious.

I moved forward so I could lean out and see how bad the fire was. I saw flames pouring out of the bottom of the fuel cell. The flames climbed up past the turbine engine and back along the tail boom.

The magnesium in the airframe was igniting.

We were running out of time.

I leaned back into the ship. A calm came over me. I relaxed into the nylon jump seat and continued firing out the door. I expected an explosion or a crash at any second.

We were all dead men.

The flames entered the cabin. My left hand and arm were being licked by the fire. I felt no pain. My flame-resistant Nomex flight suit and gloves were holding up.

I kept firing my M-60, hoping to take as many enemies with me as possible.

Smitty launched all the rockets in rapid-fire salvos to prevent additional explosions if we crashed or the inflight fire caused rockets to explode.

Lt. Cooper was leaning over to the left and not moving. I thought he had been hit.

Smitty just kept flying, putting distance between us and the dinks. He was buying lifesaving time for the rescue ship to pick us up before the enemy could kill us.

As the distance between us and the enemy increased, so did the odds of an inflight explosion.

I thought we must be quite a sight against the early evening sky in that godforsaken place. The flames enveloped our aircraft. Our rockets were launched in rapid succession. Bremer and I were putting down a steady stream of M-60 fire.

Smitty kept flying.

It had probably only been a few minutes, which seemed like an eternity since we took our first enemy hits.

Smitty brought the ship down in a rice paddy. It was a hard landing. The flames flared up and consumed the ship. The fire blocked the door gunners' exits. We had seconds to get out.

The ship was leaning to the left in the mud. The main rotor blades were dipping dangerously low on the left side.

The co-pilot was not moving.

I expected the ship to explode at any second. We were out of time.

Bremer jumped out through the flames with his 45 caliber grease gun and opened the co-pilot door. The co-pilot was not injured. The shock had immobilized him. Bremer threw back the pilot seat side armor shield, unbuckled the co-pilot, and dragged him out of the aircraft.

The Lieutenant snapped out of it.

Before I could pull back Smitty's side seat plate, he had thrown it back with some unknown physical ability. He bailed out of the ship.

I tried to get a handle on my M-60 with a belt of ammo.

Everything was tangled.

I had no time. I had to get out. A massive fuel cell explosion was imminent.

I grabbed my M-16 and a couple of magazines.

I dove through the flames and landed on my feet next to the ship.

I focused on the main rotor blades and judged their height.

I got as low as I could to avoid decapitation.

I escaped under the blade sweep, hoping the ship did not settle lower in the mud.

All four of us gathered about 100 feet away from the ship.

Flames completely consumed her.

The ammunition and grenades started cooking off and exploding. We moved back further.

All daylight was gone. The only light was 523 burning.

The Cobra arrived first. They could not recognize us in the dark. We appeared as figures, slightly illuminated by the fire. The Cobra thought we were the enemy.

The lethal gunship stopped and lowered its nose, aiming its mini-guns and rockets at us.

We frantically started waving our arms.

Smitty started yelling on his emergency radio.

The Cobra realized we were the friendlies.

They moved to a hover above us, slowing rotating 360 degrees, like an enraged mother protecting her young.

Any attempt by the enemy to advance on our position would have been met by massive lethal force.

The rescue ship arrived minutes later and came to a three to four-foot hover about 50 feet away.

Bremer and I ran for our lives to the waiting ship.

I easily jumped from the muddy rice paddy to the hovering ship's cabin deck. Bremer was right behind me and crashed on top of me. The two pilots did the same.

As the rescue ship climbed out, 523 exploded in a massive fireball.

We cheated Death again.

She was a great ship. Many times, after being shot up, she got us home safely. Finally, she just ran out of luck.

Thank you, 523.

I mentally patted her tail boom.

After Bremer and I returned to the Stinger hooch, we related our experience to the other Stingers.

The mood was reticent as we all reflected on our mortality and prospects of getting out of the war alive.

I found a felt pen.

All Stingers watched in silence as I walked over to the Stinger monument and entered "523" on the record of Lost Aircraft.

Allen back at the hooch after losing 523

CHAPTER 23 MASSIVE ENEMY FORCE

They grounded me for a few days. I had no ship. I thought I might get to go to Tan Son Nhut to pick up a brand new ship. Fate had other plans.

About mid-day, the field phone rang in the Stinger hooch.

I picked up.

It was Operations. They needed a crew chief on the flight line immediately. Sgt. McKinley had been hit. They ordered me to take over his ship as the replacement crew chief.

The truck was waiting for me. We arrived at Aircraft 470 in minutes. The door gunner was SP4 Mike Owens. He was cleaning McKinley's blood off the ship's deck.

Owens said a 30 caliber bullet had come up through the deck and hit McKinley in the butt. The bullet trajectory traveled up his body into the chest area. He was in critical condition.

McKinley was wearing a chicken plate on his chest. There was no chicken plate under the seat. Had there been a plate under the seat, it may have stopped the bullet.

The Aircraft Commanders for 470 and our wing ship arrived from Operations. The 470 AC said we were going back to the area where McKinley was shot. I had never flown with this AC before.

I geared up and was about to sit down on the blood-soaked seat when I noticed the bullet hole. I felt vulnerable with no seat plate while sitting on a very recent bullet hole.

The team of Stinger ships lifted off and headed in harm's way.

We flew for about thirty minutes when I saw many uniformed soldiers standing at attention in a company size formation.

I thought, what infantry commander would call his company into a formation in the middle of unsecured territory?

The AC was silent. No one said anything about the very unusual sight. There were over a hundred soldiers in the formation. I thought I was hallucinating. Didn't anybody see what I was seeing?

We got closer. The uniforms were not familiar.

Holy shit, they're NVA! I screamed at the AC, "They're dinks!"

The AC was silent. The wing ship was silent. The dinks did not break formation.

I thought, shit, I need to open fire and engage a massive force of NVA soldiers.

No one still had said anything. Everyone was in shock. The dinks still had not broken formation.

I opened fire on over 100 armed, trained enemy soldiers while sitting in the open doorway of a helicopter at an altitude of about two hundred feet.

Everything broke loose.

The dinks scattered as the dead and wounded that I was hitting were falling around them.

The AC finally came to life and rolled in with rockets.

The wing ship did the same.

We started taking hits.

The gunner and I returned fire.

Americal grunts were called in to sweep the area.

We took several hits, so the AC elected to get us home while we could. A crash landing in that area on top of a lot of angry NVA would not have gone well.

A couple of days later, the Stars and Stripes reported at least nine NVA were killed by the Stinger gunships. Over twenty were killed by Americal infantry.

I think it was an Americal reporter who switched the numbers in the news report. We were the first to discover the enemy and the first to engage. Americal showed up well after the enemy had dispersed.

At least Americal gave the Stingers credit for nine kills.

They recommended the Stinger crews for the Distinguished Flying Cross.

We ended up with the Air Medal for Valor, a step below the DFC.

I was on a run. The previous week, they had awarded me the Purple Heart for minor wounds incurred when 523 was shot down.

The important thing, I was still alive and physically in decent shape.

I would not fully realize the psychological damage until years after the war.

CHAPTER 24 SERGEANT CURTIS MCKINLEY

We visited Curtis twice in the base hospital. The first time he was in terrible shape with a lot of pain.

The second time, he was better. He was sitting up and said he felt good. We were very relieved. We previously thought his survival chances were near zero.

I saw an attractive Army nurse on the opposite side of the treatment bay. I walked over and introduced myself. Her name badge identified her as "Anderson".

I asked her about Curtis' condition. She negatively shook her head and said the prognosis was poor. I asked why he appeared to have improved. She replied, "Morphine".

The news visibly saddened me.

We started talking about Curtis' wound and the medical ramifications. She said I could call her "Julie". For the first time in months, I was having a conversation with a beautiful young American woman. She seemed to show mutual interest.

I felt a connection with Julie. Besides the physical attraction, Julie shared the war's horrors.

The guys were ready to go. I thanked Julie for talking to me. I asked her how much time she had left on her tour. She said she had a few months left. I wished her well. She touched my arm and said, "Good luck to you, soldier."

Sergeant Curtis McKinley died two days later. It was the end of an era for the Stingers. McKinley had been the heart and soul.

Later that night in the Stinger hooch, McKinley's door gunner, Owens, asked me for my felt pen.

As we all watched in silence, Owens walked over to the

memorial wall. He wrote in the name "Sgt. Curtis McKinley", on the Killed in Action plate.

At the memorial service, it was difficult to step forward and salute Curtis' flight helmet. A lot of history had been experienced in just a few months.

I learned Curtis was short, with only a month left on his tour. He had volunteered to come back to Vietnam again after his leave at home.

Instead of the gunships, he wanted to fly in unarmed Dustoff medivac choppers for his next tour. Curtis' intentions left a lot open to interpretation. He served as leader of the Stinger enlisted cadre, accumulated dozens of enemy kills and kept the fire alive in the Stingers to accumulate body count.

Flying Dustoff would be the complete opposite. Instead of the lethal armament of a gunship, he would only have an M-16 for defense. Instead of killing people, he would risk his life every day to save American and ARVN soldiers.

About two weeks later, we received a letter from Curtis's mother. She wanted to reach out to us because she knew how important the Stingers were to her dead son.

She described the funeral and military burial ceremony. She said Curtis' grave was covered with flowers and wreaths. One wreath had a banner with the platoon's name, "Stingers".

Mrs. McKinley wrote there was a terrible storm that night. The next day, the family went back to Curtis' grave to inspect the flowers and wreaths. She said almost everything had blown away. The only remaining decoration lying on Curtis' grave was the Stinger wreath.

They awarded Staff Sergeant Curtis McKinley, the Distinguished Flying Cross, posthumously.

His hometown renamed a city park in his honor.

The Stingers changed after Curtis' death.

CHAPTER 25 ADDICTION AND BREAKDOWN

Alcohol and drug use became prevalent in the 116th AHC at that stage of the war. The flight platoons, particularly the Stingers, were the exceptions. Perhaps our addiction was adrenaline.

All personnel could readily purchase alcohol at the PX. Soldiers who were inclined could purchase drugs practically everywhere.

There was a grunt company based next to our area. Many grunts were high after they came out of the bush. Those guys had access to any kind of drug. So, there was a fair amount of commerce between the grunts and our support personnel.

The most shocking source of heroin was the hooch maids.

One day, I walked over to the orderly room to see an old acquaintance who worked there. His name was Andy Benson. As I approached the front of the orderly room building, I noticed several small vials lying in the sand.

This was an area where the deuce and a half (2 ½ ton) trucks would load and unload hooch maids.

It was still early morning, so the hooch maids had just arrived for the day. When they jumped off the large truck, vials of heroin apparently fell out of their hair.

As I looked closer, I discovered there were dozens of tiny vials in the sand, filled with the whitish powder.

The MPs (Military Police) supposedly searched the hooch maids at the main gate before boarding the trucks.

Occasionally, 116th soldiers would be seen walking around in a heroin-induced daze.

The heroin was allegedly pure, with little or no dilution with other additives. I heard the heroin was powerful and very cheap.

There were hostile incidents involving drug transactions in the company area. Any confrontation was extremely dangerous. Everyone was armed, and some were accustomed to killing.

The Company Commander and First Sergeant positions were revolving doors. Americal was frequently firing 116th AHC leaders.

The constant leadership turnover sabotaged stability and consistent management practice.

The aircraft maintenance platoon was always pressed to the limit to keep the aircraft flying. Repairing daily battle damage and routine maintenance required 24-hour workdays, seven days a week.

The environment in the maintenance hangar was stressful and plagued by low morale. These factors contributed to high substance abuse.

Flight personnel were concerned about the quality of maintenance work being performed on our aircraft. In retrospect, I don't think any of us realized the massive scope of the substance abuse problem. Our primary focus was surviving missions and enemy rocket attacks.

The turmoil in management and leadership resulted in little attention being directed to the growing alcohol and drug problem. Soldiers were dropping out and spending days in hiding spots, getting high.

The new First Sergeant and CO called the second mandatory formation we had in a year. The purpose was to do a headcount and try to reconcile the personnel number the company was recording in its Morning Report with an actual count of live bodies.

I remember one guy who had been seen just wandering around for weeks. His skin was severely yellowed by heroin-induced liver failure.

A few minutes before the mandatory formation was called to attention, a platoon sergeant had to drag the yellow skin guy out of a bunker. He was in a heroin stupor. They propped the addict up in the formation so he could be counted.

I often wondered what happened to those guys that were addicted to pure heroin. I didn't see how these non-functioning addicts could even be processed for the return home.

Venereal disease (VD) was also rampant in Vietnam. Condoms were not readily available. It didn't matter, anyway. Midterm combat soldiers had the mindset they were already dead, so "it don't mean nuthin".

There were a couple of sexually transmitted diseases in Vietnam that were rumored to be fatal.

Perhaps the heroin addicts and terminal VD patients were sent to the same secret facility that was rumored to exist off the coast of Japan.

The rumor suggested they were reported as missing to their next of kin, never to be seen again. This is pure hearsay, with no evidence I am aware of.

Morale was at its lowest point in early 1971. America was protesting louder than ever about the war. The rumors intensified about America's withdrawal from the war.

Speculation was rampant about guys getting early rotation dates, shaving weeks off their one-year tour. Few soldiers wanted to be there.

It made even less sense to continue flying missions and risking the rest of our lives for a war that was ending without a clear victory.

The ARVNs had not shown they could take over.

Vietnamization was a joke.

The only issue that justified us being exposed to lethal enemy encounters was protecting the lives of the infantry guys and our slick crews.

November 21, 1970 - *CBS broadcast a report showing troops of the 1st Cavalry Division at a firebase smoking marijuana through a shotgun barrel.*

May 7, 1971 - *Americal Daily Staff Journal - 23d MP, MAJ Pike : Song Ice and Laundry Shop at Duc Pho raided by U.S. Army Criminal Investigation Division, QN Police, and ARVN CID in a narcotics raid. Confiscated: 758 units 4-gram plastic vials suspected heroin, 917 units 2-gram vials suspected heroin, 1 unit 20 gram stick opium, 1 unit 8 gram bottle liquid opium, 4 units 4-gram plastic packs heroin, 6 units heroin inhalers, 3 units 1/2 gram opium balls, 68 units morphine syrettes, 511 units Binoctal tablets, 102 units Immenoctal tablets, 1,020 units 4" marijuana cigarettes, 500 grams manicured marijuana. Total weight of heroin: 9 lbs, 82 grams. Also confiscated 1,000,650 piasters Vietnamese currency. Arrested: Hanh, Ngyen Thi, female 42 yrs, owner; Dinh, Cao, male, 18 yrs, Duc Pho PF. Info given to USARV and XXIV Corps.*

CHAPTER 26 SPRAY MISSION AGENT ORANGE

I took over Curtis McKinley's ship, 470, after Curtis died of his wounds. For some time, I did not feel the same bond with 470 that I had with 523. It just seemed it was Curtis' ship. I guess I felt Curtis' presence as I occupied his seat with the hole from the bullet that killed him.

It is now well known the U.S. Army sprayed vast amounts of defoliant on the Vietnamese countryside. The intent was to kill vegetation the enemy used as camouflage and rice paddies that produced food.

The defoliant was fast-acting and effective in destroying any kind of plant life.

The airmen responsible for disbursing the herbicide with aircraft did not know the chemical cocktail as "Agent Orange". The soldiers also did not know the long-term effects the chemicals had on the human body.

It was not uncommon for an airman to be soaked in the chemical during aerial spray operations.

The risk of chemical contamination was way down on the priority list of threats airmen faced in Vietnam. It was almost laughable to be concerned about ailments that might arise years after the soldier returned home.

The top concern was to survive to see the next day.

Stingers were occasionally assigned the duty to protect Huey slicks rigged as spray ships. It was not one of my favorite missions, however, there were some exhilarating experiences.

One Stinger would follow the spray ship as it flew low over the countryside and sprayed the defoliant. The spray ship had two spray booms that extended about ten feet on both sides of the cabin. They secured the chemical tank in the cargo area of the cabin. As the spray ship flew along doing its duty, some chemical would ride the air stream and soak the following gunship gunners who were sitting in the open doorways.

The upside was since it rained every day and the chemical was soluble, we just had to fly through a rain shower to get rinsed off.

Sometimes, a VC would pop up and take a few shots at the spray ship or the guardian gunship. We would instantly swoop down on the offender and kill them with M-60 fire.

One mission did not follow the routine. We were following a spray ship one day, getting soaked with chemicals, when multiple streams of tracers started streaking by the spray ship and then our gunship. I could hear several weapons firing.

There was a loud bang as a bullet broke through the cabin floor between me and the gunner. Shrapnel from the metal floor with ammo links and brass cartridges lying on the floor were propelled throughout the cabin by the force of the bullet.

The spray ship gunners were closer to the enemy. They opened up M-60 fire on both sides of the ship. We could see the slick gunners hitting their marks as dinks were hit and fell.

It was a special treat for the slick gunners since they rarely got kills. The spray ship broke off and retreated from the area.

Our gunship took over. We had no wing ship, so we were on our own against about twenty VC.

The AC lined up the ship and fired multiple salvos of rockets into the concentrated area where the dinks were firing from.

The gunner and I fired the M-60s into the group of dinks with deadly accuracy. We made three left-hand orbits around the enemy location, continually firing rockets and M-60s. It appeared we had neutralized most dinks, so Owens, the gunner, got in some target practice with the M-79 grenade launcher. Bodies and debris exploded upward as the grenades impacted.

Satisfied we had annihilated the threat, the AC broke off the attack and we rejoined the spray ship for some more defoliation.

Just another day at the office.

Allen having a C-ration lunch while the spray ship reloads with defoliant

Huey Spray Ship

CHAPTER 27 SP4 TONY BREMER - SILVER STAR

Bremer caught a mission on another Stinger gunship as the door gunner. The Crew Chief was our black guy, SP5 Alvin Jones. As far as I know, Bremer's prejudice had not diminished, even though Jones was well-liked by everyone else.

There were two Stinger gunships on an LZ insertion mission. The guns went into the LZ area first, as usual, to draw out and suppress enemy resistance.

The area was red hot. Both ships started taking intense enemy fire. All four gunners on the two ships were hanging out over the rocket pods, returning fire to save their ships.

The Aircraft Commanders were firing rockets as the ships orbited the enemy positions. Jones and Bremer's ship banked hard left to avoid a stream of deadly enemy tracers.

Jones was leaning out of the ship, returning fire when he experienced a weapon jam.

The sudden recoil stoppage caused Jones to fall forward, accelerated by the steep bank. He was not wearing a seat belt or monkey strap.

Jones fell out of the helicopter from about eighty feet AGL and slammed into a muddy rice paddy. His M-60 with a few feet of ammo belt fell next to him.

Jones was motionless.

The enemy was about 200 feet away.

The AC quickly called the other Stinger, advising of the crewman down.

Forgetting his inbred prejudice, in a flash of duty and courage, Bremer screamed on the radio, "Get me down and I'll go get 'em."

The AC coordinated with the other Stinger to provide cover while he maneuvered the ship to the ground near Jones.

Jones was still motionless.

The enemy continued firing from both sides of the LZ. The descending ship took multiple hits.

The wing Stinger rolled in on the enemy positions, laying down rocket and M-60 fire.

Before the Stinger rescue gunship touched down, Bremer jumped out and ran about 30 feet to Jones with enemy bullets hitting the mud all around him.

Bremer picked Jones up, who weighed around 180 pounds and threw him on his shoulders in a fireman's carry.

Bremer could not retrieve Jones' M-60.

He ran back to the waiting gunship with Jones unconscious, balanced on his shoulders.

Intense enemy fire continued striking near his path.

The Stinger ship took over twenty hits.

Bremer was not wounded.

Jones was still alive, but unconscious. His back had been injured in the fall.

The United States Army awarded Specialist Fourth Class Tony Bremer from Alabama, the Nation's third-highest honor for valor in combat, the Silver Star.

Bremer never mentally recovered.

The trauma of 523 crashing and burning and our miraculous escape from death just a few weeks before was difficult to bear.

Before that, he witnessed his friend, Mark Rowland, dying in his jump seat from a fatal enemy bullet.

Combined with the selfless act of courage to rescue a black man from certain death, in contradiction to his lifelong inbred prejudice, was more than Bremer could handle.

Later, Bremer approached me, seeking help. He tried to verbalize what he was feeling, but could not.

Regrettably, I offered no solace. I was consumed with my own internal conflict. It was all I could do to just hang on and do my duty.

Bremer withdrew into his own world and shut out everything and everybody.

He disappeared without notice. I am not sure if they transferred him to a medical facility or what happened to him.

By that time in the war, Stingers were not talking to each other about their daily war experiences. There were no more celebration victories for staying alive and killing dinks.

Flying missions just became a daily ritual in a constant numbing atmosphere of death.

Surviving Stingers would just disappear and go home with no farewells.

CHAPTER 28 DEAR JOHN

An issue that has plagued wartime soldiers since the beginning of a postal service and now the internet, is the "Dear John" letter.

The dreaded letter from the loved one at home announcing they had found another and would not be there when you returned from combat duty.

My friend in the orderly room, Andy Benson, was married with two young children. He carried their pictures with him and showed the photos to anybody that would look.

Andy got the "Dear John" letter. To say he was devastated would be an understatement.

After losing his family and coping with the toxic environment we were all enduring, it became too much for him to bear.

He decided there was nothing to live for. He volunteered to be a door gunner on a slick.

I pleaded with the guy not to take that course and tried to convey the immensity of danger in flying assault missions.

He would not listen.

Less than a month later, SP4 Andrew Benson was burned alive when his helicopter crashed and fell on the side where Andy was seated.

Andy was trapped inside the door gunner well.

The ensuing magnesium-fueled fire consumed the ship and Andy's trapped body.

There was a story about a guy in another unit who received the "Dear John" letter. While a helicopter was in flight idle, sitting on the ground, the heartbroken soldier walked into the tail rotor. It sliced his head and upper chest area in two.

Two Stingers received "Dear John" letters. Both had just a couple of months left in their tour.

One of the guys just became angry about everything and everybody.

The other Stinger became morose. I never heard him speak again.

The atmosphere in the Stinger hooch worsened. The overall declining morale as the war wound down and constant lethal incidents disheartened most flight personnel.

As I have pointed out, being short is the worst time for combat soldiers. You have growing hope you are going to survive. You know every new mission you fly or any incoming rocket attack could be fatal.

The growing hope is offset by growing fear.

For most, the greatest dream of all is surviving the war and being back with your loved ones.

When the soldier learns the dream has vanished, the loss is magnified.

CHAPTER 29 NIGHT MISSION - HAWK HILL

Hawk Hill was a firebase between Chu Lai and Da Nang.

Army Command would often order the Stingers to pull overnight standby duty at Hawk Hill.

Our mission was to provide immediate fire support for Americal infantry units in the area, which might get attacked, usually at night.

There were few places in Vietnam where I had been that were worse to spend the night than Hawk Hill.

Unlike the grunts that endured life in the jungles for weeks at a time, airmen were accustomed to going home at night. We could get a hot meal and drink a few brews. We slept on a bed with sheets and a pillow with a roof over our heads. We could even take a shower. It spoiled us.

The only thing Hawk Hill offered was a roof over your head, which was a sandbag-covered bunker and a few ramshackle hooches.

There was a lot of company in the bunker. Enormous rats were constant companions as they scavenged food.

One guy had his penis bitten by a rat when he took a dump in the latrine.

A mission to support infantry at night by air came with a unique set of risks. We had no night vision equipment. The only way to identify where the friendlies were was by light signals or flares.

The grunt commander could communicate with the gunship AC by radio. He could give bearings from his light signal or flare location where the enemy was and what the radius of friendlies was around the light source.

That strategy was risky. So, if we got called out on a mission at night, the tactical situation was critical.

We got a call one night at Hawk Hill from an Americal platoon. They were being attacked by a superior enemy force and were threatened with being overrun.

We scrambled our ship and lifted off to confront the enemy to save our guys on the ground.

There were no wing ships on night missions for fear of mid-air collisions in the dark.

It was pitch black with no moon. I could not see a thing. It was just a black abyss.

The AC contacted the platoon leader. The AC also opened the channel with the grunt commander so the entire crew could hear. We would all need to know where to direct fire and how to avoid our guys.

The commander advised the area was hot, and they were taking fire. He reported they had casualties. He believed the enemy was advancing from the north and west of his position.

About five minutes later, the commander reported he could hear us coming. He said he would turn on his light and oscillate the light filter from white to red.

We saw the enemy tracers first. As we approached, we searched for the commander's light.

The AC advised the grunt commander we were going dark, turning off all navigational lights.

As we flew in total darkness, searching, there was no light from the ground except a few enemy tracers probing the night sky.

The grunt commander did not answer our radio calls.

We did not know if the enemy had overrun the Americal position.

We could return fire in the areas where enemy tracers were being fired at us, but we needed to know our friendlies' position.

Finally, an unfamiliar voice came over the radio. The soldier reported the commander had been hit and was down. The soldier identified himself as the platoon sergeant. The Sergeant retrieved the commander's light and signaled us with a rotating white/red.

We saw the light.

The Sergeant directed us to attack fifty yards to the north and west of the light.

The AC rolled in on the area with rockets firing. The gunner and I aimed forward outside the ship and laid down M-60 fire.

The increasing enemy tracers streaked past us on both sides.

We were nearly invisible in the night sky. Only our M-60 bursts and rockets launching could give the enemy our precise location for a few seconds.

The sound of our turbine engine and rotor blades would only reveal our general location.

The night sky filled with tracers as the enemy blanketed the airspace to shoot us down.

Every fourth or fifth round fired was a tracer. So the number of bullets being fired at us was multiplied by the bullet tracers we could see.

The AC banked into a left orbit around the targeted area. The enemy was stationary. We could return fire every time we saw enemy tracers fired from the ground.

The AC called the Americal sergeant for a report on our firing accuracy. The Sergeant reported we were clear of the friendlies and it appeared we were hitting our intended targets.

We continued making left-hand orbits around the area where the tracers were coming from.

Owens and I laid down steady streams of M-60 fire.

The AC fired about twenty more rockets into the area where the tracers were being fired at us.

The enemy got a fix on us. We started taking hits.

Owens and I kept firing. Our M-60 barrels glowed red in the dark.

The enemy stopped firing.

The Americal sergeant called and said, "You guys got 'em! Thank god". We could hear him sigh with relief on the radio.

The grunts had four wounded and two KIA. The AC contacted Dustoff.

When Dustoff arrived, the grunts set off flares for a landing area. We circled the area, ready to repel any enemy attack.

After Dustoff lifted off with the casualties and KIA, the AC radioed the Sergeant and asked if there was anything else we could do for them.

The Sergeant responded, "No, thank you, you did plenty."

We flew back to Hawk Hill.

At first light, I inspected the battle damage on the ship. We took four hits. No critical systems were hit. We could fly back to Chu Lai on our own.

The "personal kills" metric proved to be an elusive number for crew chiefs. It was clear you had a confirmed personal kill when both you and the AC saw the dink take a fatal hit. It was impossible to know how many dinks I killed when I was spraying hooches with M-60 fire or firing into the jungle, day or night, at groups of the enemy.

I guess the "personal kills" number was just for bragging rights.

The real numbers of enemy casualties were certainly higher.

If a crew chief had thirty-five confirmed personal kills, the real number probably exceeded 100.

CHAPTER 30 YELLOW JACKET DOWN

We got a mission to fly cover for a Yellow Jacket insertion near Ky Tra fire support base, about twenty miles west of Chu Lai.

We arrived at the flight line at 0800 hours for a 0900 lift-off. We had plenty of time to do a thorough inspection of the aircraft and armament.

I found a latch broken on the right-side engine compartment cover. The damage was a "Red X" condition grounding the aircraft.

That day, we had a new AC. He was not happy about me grounding the ship for a broken latch. The issue was, if the airstream lifted the cover enough to break it off in flight, it could fly up into the main rotor blades or back into the tail rotor.

The AC did not care. He was determined to make the lift-off time.

I could ground the ship no matter what the AC wanted. It was a battle I wanted to avoid.

I ran to the maintenance hangar and grabbed the tech. I got him up to speed on the problem. The tech dropped what he was doing, commandeered a mechanic, and got the part from the supply department. We ran back to the ship.

There were thirty minutes to change the part before the scheduled lift-off. While the tech and mechanic were replacing the latch, the AC continued to bitch about the delay.

With ten minutes to spare, they completed the repair.

The tech turned to the AC and said, "It's a good thing the crew chief caught this bad latch. A similar problem occurred in another unit. The broken latch was not repaired, and the cover broke off in flight. The large cover flew back into the tail rotor. The aircraft lost anti-torque control, and the AC had to perform a hair-raising, running landing to save the ship and crew."

The AC got quiet and never said another word.

If the tail rotor fails, the ship will be uncontrollable and spin with the main rotor.

If autorotation is not an option because of an unsecured area with enemy activity, the ship must return to base.

The pilot must maintain a speed of over sixty knots.

A running landing is performed to keep the speed up during the approach so the wind vane airflow will keep the ship from spinning.

The AC hits the runway at above sixty knots, skidding along until the aircraft stops.

It is a tricky maneuver that requires a high level of pilot expertise to prevent a crash.

It is worth the time to fix the broken latch.

The Americal infantry guys loaded up on the Yellow Jacket slicks.

The pair of Stinger gunships lifted off about five minutes before the Yellow Jackets.

An Americal infantry platoon had encountered a force of about sixty NVA, two miles north of Ky Tra the day before.

The artillery unit based at Ky Tra had orders to begin a fire mission on the area around the intended LZ at 0855 hours.

Our lead Stinger AC contacted the arty (artillery) guys to confirm the fire mission had ceased when we were about ten minutes out.

The Americal artillery barrage hopefully had softened up the

NVA force.

The Stingers rolled in on the LZ with M-60 fire to ensure the enemy had been subdued before our Yellow Jacket insertion.

There was no enemy contact as we laid down machine-gun fire on two orbits around the intended LZ.

The AC called the Yellow Jackets and gave the signal for the execution of the landing.

The NVA were excellent soldiers. They had become familiar with Americal tactics. The advance artillery barrage alerted the NVA we were on the way.

They knew the gunships would prep the LZ, so they stayed camouflaged until the gunships had cleared the area and gave the go-ahead to the slicks.

The Stingers continued left-hand orbits around the LZ as the Yellow Jackets approached and landed in a staggered formation.

The last slick landed and the lead ship was about to lift off after offloading its load of grunts and supplies.

The NVA popped out of their well-concealed camouflage and opened fire.

I yelled, "Taking fire" as tracers flew by our ship. Owens and I returned fire.

The AC lined up the ship on a target and fired rockets.

The NVA started firing mortars on the slicks.

The lead slick was hit, slowing down the escape of the rest of the flight. The second ship took command and ordered the rest of the gaggle to maintain the formation to avoid collisions.

The slick gunners were spraying the LZ perimeter with M-60 fire between and over the heads of the grunts that were entrenched near the ships.

As the slicks gained altitude, the gunners sprayed the perimeter as they departed.

Overhead, the Stingers continued firing rockets and M-60s.

The grunts on the ground were targeting pockets of the enemy with small arms fire and mortars.

The enemy was being bombarded from the air by the gunships, from the ground, by the grunts and the slicks departing the area.

Soon, the enemy fire subsided.

The grunts formed a defensive perimeter around the downed Yellow Jacket.

The Stingers flew orbits overhead in a tight radius.

The downed crew and ship were well protected.

The grunts and slick crew had casualties. The Stingers stayed on station protecting the grunts and downed ship until a Dust-off arrived to pick up the casualties.

The Americal unit reported the Stingers had killed nineteen NVA.

CHAPTER 31 ARVN LZ INSERTION

The Yellow Jackets and Stingers got a mission to insert ARVN troops in a landing zone near Tam Ky, fifteen miles northwest of Chu Lai.

We expected the LZ to be "cold" (no enemy resistance), based on ARVN intelligence.

Americal opted not to order an artillery fire mission to soften up any enemy resistance. The ARVNs assured Americal the area was cold.

The mission started at Chu Lai. They loaded the ARVNs on the slicks. The Stingers took off first to recon and prep the LZ before the insertion.

It was about a twelve-minute flight to the insertion point.

The slicks were five minutes behind us. My ship 470 was lead with a wing ship.

We started left-hand orbits around the LZ.

The gunners sprayed the area with M-60 fire to draw out any enemy.

After two orbits, the LZ appeared to be cold. No enemy contact.

Our AC gave the slicks the signal to proceed.

The Stingers remained on station, orbiting the LZ.

The slicks entered the LZ and settled down to offload the troops.

Several NVA who were hidden in the underbrush around the

perimeter sprang out and started firing in a coordinated attack.

The slicks were taking hits. ARVN troops were being slaughtered.

The Stingers rolled in for the counterattack. We fired rockets towards the NVA firing positions. I killed every NVA I could see as we continued left-hand orbits.

Mike Owens, my right side gunner, pulled his ammo can over to my side and started firing his M-60 out my doorway to concentrate our firepower.

Command ordered the slicks to escape and abandon the mission.

The Stingers remained on station, killing NVA.

We intended to wipe out the NVA unit.

After the slicks exited the area, the NVA turned their full attention on the Stingers. We started receiving increased small arms and machine gunfire.

Bullets blasted through the deck, just missing Owens and me. We responded with increased ferocity.

I watched my bullets rip apart two NVA in one position.

Owens started racking up kill count from my side of the ship.

We fired all thirty-eight rockets in the NVA positions.

The wing ship did the same for a total of 76 rockets.

The ground looked like a mini B-52 run.

We counted 28 dead NVA.

The enemy paid dearly for the surprise attack.

We heard there was an investigation into how the mission was leaked. The ARVNs buried it.

At least no Hornet crewmen were injured or killed. Our ships, however, took severe beatings.

It was a couple of days before I got 470 back from maintenance.

Our collective opinion of the ARVNs was pretty low. The leaked mission had to come from an NVA spy posing as an ARVN. It would be prudent to stay vigilant any time we had missions with the Army of the Republic of Vietnam troops.

About a week later, we received information from Intelligence. The NVA had posted a bounty on any Stinger, Dead or Alive. The information had an NVA bounty flyer attached portraying the Stinger gunship nose art.

We were proud the NVA acknowledged us as a serious threat.

Our suspicions about NVA infiltration into the ARVN military and Saigon government would soon be confirmed.

New York Times- Washington, Oct. 18, 1970 -*The Central Intelligence Agency has told President Nixon that the Vietnamese Communists have infiltrated over 30,000 agents into the South Vietnamese Government in an apparatus that has been virtually impossible to destroy.*

Because of this, the C.I.A. reported, as United States troop withdrawals proceed, a resurgence of Communist strength in South Vietnam can be expected.

The report to Mr. Nixon said the secret Communist agents included an aide to President Nguyen Van Thieu of South Vietnam, a former province chief, and top officials of the police and military intelligence.

Confirmation by Officials

While the study is not addressed specifically to the question of the President's war policy, officials of the United States Government who have read it say that it raises questions about a key aspect of this policy—Vietnamization, or gradually giving the South Vietnamese the main burden of defending themselves against the Vietcong and North Vietnamese and thus allowing American troop withdrawals.

The scale of communist subterfuge was so massive, the chances of South Vietnam surviving were negligible.

The declining number of U.S. troops still in Vietnam from 1970 to 1972 were in great peril, as the Communists hastened the collapse of South Vietnam from within.

Stinger 470 Nose Art

CHAPTER 32 DONUT DOLLY RUN

After months of seeing nothing but slant-eyed Vietnamese women, most soldiers yearned to be in the company of "round-eyed" American women.

We heard of the Red Cross, "Donut Dollies" who visited the troops in Vietnam. The odds were slim the aircrews might get the chance to revel in the presence of American "Dollies".

We figured we had about the same odds of enjoying a "Dolly" visit as seeing a Bob Hope show.

If you were not in Saigon, Da Nang, or Cam Rahn Bay, which were secure and about as safe as you could be in a war zone, you had almost no chance of attending Bob Hope's show with his usual beautiful co-hosts.

Unbeknown to us, some Red Cross girls did venture into the wilds of Vietnam and visit front-line posts.

We were about to embark on a search and kill mission one day when we got word we would carry a few passengers.

That was strange since we had no additional passenger seats and we would certainly get shot at that day.

Our passengers showed up, and I became unhinged at the sight.

Three young, cute American girls in tight little Red Cross outfits jumped out of the jeep. I thought I must have taken a bullet to the head, and I was in heaven.

Our mission was to drop the girls at Hawk Hill.

Considering Hawk Hill often took incoming enemy artillery, I would say the girls were incredibly brave or just unaware of the danger.

The girls had to sit on the deck amid the cans of grenades, assorted weapons, M-60 ammo belts, and two gawking door gunners.

Considering we were in combat mode, about to go out to be shot at, coupled with being freaked out by cute American girls; any suave and debonair mannerisms escaped us.

The girls were uneasy.

When we crossed the perimeter and prepared rocket circuit breakers and M-60 machine gun safeties for a life or death battle, the girls visibly became shaken.

I noticed one girl looking at the bullet holes in the deck where enemy bullets had struck through.

The reality showed in their faces. This was not a pleasure cruise.

We got the girls to Hawk Hill unscathed.

The sight of the half-naked, filthy Hawk Hill combatants probably unsettled them more.

In retrospect, I must salute those young women who braved the combat environment to bring a little light into the dark, wicked world the soldiers lived in.

We would have gladly brought the girls back to Chu Lai, but we had to go kill dinks that day.

I am sure one of the many helicopter crews visiting Hawk Hill jumped at the chance to ferry the girls back to relative safety.

CHAPTER 33 CW2 LOU BALDASSARO

Lou was short.

He had 100 kills.

He had achieved his goal of 100 kills with Stinger rockets.

Americal refused to recognize Lou's last kill because an Americal unit failed to confirm.

Lou became infuriated.

He fought his way through flight school as a very young enlisted candidate. He excelled in the Stingers as a pilot and leader. He intended to stay in the Army as a career. This was a guy who set objectives and refused to fail.

The Vietnam War was about enemy body count.

In Lou's mind, if he could achieve his goal of 100 kills, he would consider his time in Vietnam as successful, no matter what happened on the macroscale.

As a career officer, they would record his official body count number in his efficiency record.

Lou got down to just a few days before his time to go home.

They usually grounded pilots when they were within a week or two of rotation.

Lou insisted on continuing to fly missions until he got his 100th kill.

Lou got his chance when he found a lone NVA. He promptly hit the enemy with a 17-pound high explosive rocket.

The NVA's body fragmented into gory pieces.

There was no Americal unit in the area to confirm the kill.

Lou then did the unthinkable.

He landed the gunship and told the crew chief and gunner to cover him.

He got out of the ship and walked over to the largest fragment of the NVA he just killed.

He then dragged the bloody mass back to the ship and tied the partial body to the nose of the ship.

Lou climbed back into the gunship and lifted off for the trip to Americal Headquarters in Chu Lai.

He landed on the Commanding General's pad in front of Americal.

He shut down the helicopter, walked through the building's main entrance, and stepped up to the receptionist's desk.

Lou then loudly asked, "Who can I get to come out to the pad and verify my 100th kill."

What happened after that is speculation.

Most personnel at Americal Headquarters had probably never seen a dead dink.

I would think the image of a blood-dripping partial body strapped to the nose of a gunship sitting on the General's pad would have some repercussions.

Since Americal was not a big Hornets fan, I surmise that whether Lou got his 100th kill confirmed was moot at that point.

I never saw Lou again.

Everyone was tight-lipped if they knew anything.

I hope it worked out for him.

He may have been the greatest Stinger that ever served.

CHAPTER 34 JESUS NUT

Aircraft 470 came due for its 100-hour inspection. The ship came out of maintenance a few days later. We got a search and kill mission.

With all the issues regarding substance abuse and the rumors that our maintenance guys were abusers, I was concerned about the new 100-hour inspection.

All major components are disassembled, inspected, serviced, and reassembled.

I went to the flight line an hour earlier than the rest of the crew and started my detailed inspection. I started at the nose with the avionics compartment and worked back. I checked everything I could access. I covered every inch of the aircraft until I got to the transmission/main rotor pylon.

Working from the bottom up, I got to the top of the mast. I started checking the stabilizer bar, swashplate, push-pull tubes, and rotor head at the top of the assembly.

Everything was OK. All connecting rods were tight.

For the first time, the thought of checking the Jesus Nut flashed in my mind.

I almost dismissed it. What is there to check? They tightened the Jesus Nut with a torque wrench that had a four-foot handle to leverage the high load that was required to tighten it to specification.

It seemed ludicrous to check it with my fingers.

I climbed down from the mast to the ground.

I laughed it off again.

But something kept nagging at me.

My instinct said, "Go ahead, and check the damn thing."

I succumbed to instinct and climbed back to the top of the mast and put my hand and fingers around the large nut.

I applied pressure to turn the Nut.

The Nut easily turned.

My mind could not process the unthinkable for a couple of minutes.

I had to hand tighten and loosen the nut several times before I realized we never would have completed our search and kill mission that day.

Enemy bullets would not have killed the entire crew.

Our demise would be because of the negligence of our maintenance guy, who was supposed to torque down the Jesus Nut.

The Jesus Nut holds the entire main rotor assembly on top of the mast. The mast is the shaft that connects to the transmission. If the Jesus Nut comes off in flight, the entire main rotor assembly comes off.

The fuselage containing the crew would have fallen from the sky like a lead brick. With any appreciable altitude, survivability would have been impossible.

I gathered my wits and climbed down to the ground.

I marched to the maintenance hangar and located the maintenance tech supervisor.

I did not want to make a scene. In the most controlled manner I could muster, I conveyed to the supervisor that someone failed to torque down my goddamned Jesus Nut.

The supervisor had as much difficulty comprehending as I

did.

First, he asked, "Did you take a torque wrench and check the load?"

I replied, "No, I loosened it with my hand."

The supervisor looked at me in disbelief.

I invited him to come with me to the ship and see for himself.

We were both silent as we walked across the tarmac to the flight line.

The supervisor climbed up the pylon assembly to the top of the mast. He grabbed the Jesus Nut and witnessed the critical component turn in his hand.

He was visibly shaken. He fumbled in his attempt to climb down from the ship.

His face was contorted and pale.

He did not say a word as he walked back to the maintenance hangar.

He came back shortly with the four-foot-long torque wrench. He climbed back up the mast and torqued the Nut to specification.

He climbed down. As he walked away, all he could say was, "I'm sorry. Thank God you caught it."

As I reflected on that incident, I realized the irony.

I had survived countless enemy bullets, shot down in flames, and nearly blown up by enemy rockets.

That day, a loose Jesus Nut would have killed our entire crew.

The Jesus Nut

CHAPTER 35 FIRE SUPPORT BASE MARYANN

The Stingers had flown several missions in the FSB Maryann area. The Base was strategically located on the western border of South Vietnam, near branches of the Ho Chi Minh trail. The purpose of the FSB was to interdict the movement of enemy troops and supplies.

U.S. Command had scheduled FSB Maryann to be handed over to the ARVNs.

For months, enemy activity had been minimal in the area. The low military activity and the imminent transfer to the ARVNs had rendered the Americal infantry unit based there, rather complacent.

My ship, Stinger 470, and a wing ship caught a mission to Maryann to conduct a search and kill operation.

Because of the lack of recent activity, we assumed the mission was more of a reconnaissance patrol.

We arrived in the morning and started the search.

We contacted scattered enemy troops and engaged.

Curiously, the NVA were not aggressive. They returned our M-60 fire but did not persist.

We continued to find NVA, attack, and then chase survivors.

There were a few encounters like that before we ran low on fuel and had to break off.

Our lead AC reported the skirmishes to Operations. He noted the NVA did not seem willing to engage. It appeared they were there in force but were not ready to attack.

Ground sensors hidden in the jungle surrounding the Base

continued to pick up movement. The FSB was in a straddle with overlooking hills on three sides. The enemy had several vantage points to monitor the FSB from higher ground.

Later in February 1971, further northwest in Laos, one of the most intense battles of the Vietnam War, would be fought by the ARVNS supported by U.S. air assets. The operation would be called Lam Son 719. It would become clear the North Vietnamese Army had moved much of their entire force to the region.

During that time, even with the largest NVA force ever assembled less than 200 miles away, reports suggest the Maryanne Base security was lax.

The perimeter was not continuously protected. Trip flares set in the perimeter wires were not reset after being tripped by the few enemy probes or the large CH-47 Chinook supply helicopter rotor wash.

The defenses were under-manned. There was a lack of supervision and leadership at the unit command level and higher.

By late March 1971, the massive NVA force had defeated the ARVN troops in the Lam Son 719 operation. On March 28, 1971, in the early morning, Viet Cong sappers attacked FSB Maryann in force. An intense enemy mortar attack began at 0230 hours.

A force of about fifty sappers attacked in squads of three to six men. The sappers were armed with AK-47s, rocket-propelled grenades (RPGs), satchel charges, hand grenades, and CS gas grenades.

The enemy surveillance had identified the sapper targets from surrounding hilltops in the previous months.

Artillery pieces, the unguarded Battalion Ops Center, company command post, and many perimeter bunkers were destroyed or heavily damaged by explosives.

Thirty-three American soldiers were killed, including several commanders. Eighty-three were wounded.

Army Chief of Staff General William Westmoreland assumed command of the investigation of the attack.

The General cited clear dereliction of duty, lax behavior, and failure of officer leadership as the reasons for the attack.

The Army brought charges against six officers, including the Americal 23rd Division Commander and Assistant Commander.

CHAPTER 36 FIRST FEMALE ENEMY KILL

Operations scheduled six Yellow Jacket slicks and two Stinger gunships to support an Americal ground mission.

My ship, 470, would be the lead gunship. The AC would be CW2 Tommy Martin with a new guy co-pilot CW-1 John Franklin and Door Gunner, SP4 Mike Owens.

The insertion point was near a village about sixty miles northwest of Chu Lai.

The slicks could make the round trip without refueling, but the Stingers would have to refuel in the wilderness somewhere.

Gunship fuel capacity was less than slicks. We also required more fuel for the additional time of loitering over the target area and potential enemy contact.

Americal placed a large fuel bladder full of JP4 in a tiny village, several miles from the target LZ. We would have to hand pump the fuel, extending our exposure time on the ground.

As usual, the Stingers led the armada into battle. As we approached the intended LZ just outside a small village, we began searching for the enemy.

There were no reports of enemy activity in the village, but we were less than confident in Americal intelligence gathering.

As we started our left-hand orbit around the LZ, we took fire from the village.

I spotted a female dink firing at us.

The AC banked around the target.

The dink continued firing at us with an AK-47.

I returned fire with my M-60.

I saw her recoil backward as my 7.62 rounds hit her body.

The AC descended closer to the dying dink.

I could see her facial expression.

I finished her with a quick burst of lethal fire.

The wing ship had broken off to search the rest of the village. No other enemy soldiers were spotted.

Our AC gave the slicks the signal to insert the troops.

After the insertion, the slicks departed the area for the return trip back to base.

The Stingers loitered in the area to support the infantry if needed.

We were burning about 80 gallons of fuel per hour.

The infantry commander released us about forty minutes later. We could barely get back to base without refueling with no reserve.

The AC elected to find the fuel bladder in the wilderness and refuel.

We found the bladder near a small group of hooches.

It was a very uneasy feeling, landing in an unsecured remote spot.

Our ship would land first. The wing ship would orbit above us to cover.

We would switch roles after we refueled.

It took much longer to refuel with a hand pump.

We took turns working the pump.

While I was getting a break, an old woman approached me from one of the hooches.

We all readied our weapons.

The old woman smiled, exposing black teeth from years of chewing the addictive Betel nut.

She gestured for me to follow her.

I was reluctant, be she persisted.

Out of curiosity, I followed her.

I asked Owens to cover me.

I cautiously followed the old lady with my M-16 poised for quick-fire and safety off.

She led me to a nearby hooch.

I looked inside the hooch.

There were five or six very young girls sitting or lying on floor pallets. The oldest girl may have been twelve years old.

It appeared the old woman was running a brothel staffed by children.

I declined the offer. Owens and the pilots also declined.

A kind of sick feeling crept over me.

It may have been from shooting the female soldier and then seeing her facial expression as I fired the final shots into her body.

Then seeing very young girls being prostituted in this faraway foreign place.

This country was a nightmare, and we were major participants.

The feeling of patriotism in fighting for America made little sense anymore.

I believe the sliver of emotion I was feeling may have been the last trace of humanity I had left.

Or it was the resurgence of my humanity.

That was troublesome.

I could not be a Stinger if I became humane again.

CHAPTER 37 LRRP TEAM EXTRACTION

LRRP is the acronym for Long Range Reconnaissance Patrol. These elite forces were small, well-armed units that patrolled deep in enemy-held territory. The specially trained combatants were known for achieving high enemy body counts.

LRRP operatives required high-calorie food to sustain their rigorous missions. LRRP rations were meals in sealed plastic pouches that were boiled in water and served hot. Without a doubt, LRRP rations were the best Army fare that could be found in Vietnam.

It was a rare treat to abscond with a supply of LRRP rations.

At the end of a mission, an LRRP Team would contact the base and request extraction. LRRP extractions would usually be in high-risk locations, so the mission required a pair of Stinger gunships to provide cover.

We got the call to accompany one of our Yellow Jacket slicks to extract an LRRP unit about fifteen miles northwest of FSB Maryann.

The pickup zone (PZ) was a hole in the heavy jungle with high trees. The slick would have to achieve a dangerous 100-foot hover and drop straight down into the hole.

Our AC contacted the LRRPs and requested their position relative to the extraction point.

As we neared the PZ, the LRRPs reported intense enemy contact.

Our Stinger AC requested a bearing to the enemy from the LRRPs' position.

The LRRP Team Leader responded the enemy was firing from

positions about 10 degrees to 200 degrees. The Team Leader estimated forty NVA in the enemy force.

The LRRPs were in danger of being outflanked. It would be suicide for the Yellow Jacket to descend into the opening in the jungle. The enemy had to be eliminated.

Our AC contacted the LRRP Team Leader to confirm their position. He told the Team Leader to keep his men low near the boundary of the PZ.

The Stingers rolled in on the enemy. We were taking intermittent fire as we strafed the areas around the PZ, opposite the LRRPs.

The dense jungle impeded the enemy's ability to get clear shots at the two Stinger warbirds.

The LRRPs reinforced their flank positions near the boundary of the PZ.

Owens fired M-79 grenades into the jungle areas, concealing the enemy.

The dinks got lucky.

Bullets blasted through the pilots' radio console, knocking out our ship's communications. Another bullet hit the bottom of the co-pilot's armored seat. The co-pilot instantly yelled out in shock.

He quickly acknowledged he was not injured, just rattled.

The AC immediately contacted the LRRP commander and wing Stinger by handheld radio.

The two Stinger ships set up rocket runs, flying over the LRRPS and aiming at the opposite sides of the PZ.

We worked our way around the PZ, opposite the LRRPS, laying down M-60 fire, M-79 grenades, frag grenades, and 17 pound high explosive rockets.

The incoming fire slowed down as the enemy attempted to take cover from the massive attack from the sky.

The LRRPs reported the enemy was being forced towards them, allowing the LRRPs to pick off enemy soldiers one by one.

The assault lasted about fifteen minutes. The LRRPs reported the enemy had been killed or retreated into the jungle.

The PZ was clear.

The Yellow Jacket came to a hover above the PZ and slowly descended into the hole. As the Yellow Jacket touched down, the LRRPs rushed into the PZ and piled on the waiting slick.

The Yellow Jacket AC struggled to achieve enough lift to climb out of the hole with additional weight and no wind.

The Stingers continued orbiting, laying down M-60 fire to suppress any enemy that might take a shot at the vulnerable slick as it struggled to escape the hole in the jungle.

The Yellow Jacket cleared the hole.

The Yellow Jacket slick, loaded with LRRPs and its Stinger protection ships, formed a small formation and headed southeast to Chu Lai.

Later that night, the Yellow Jacket crew came to the Stinger hooch and delivered a gift from the LRRPs.

A case of LRRP rations.

CHAPTER 38 VENGEANCE

A Stinger ship was shot down and crashed in the early morning. A large caliber enemy bullet killed a new door gunner, striking him in the chest. He died instantly.

His name was SP4 William Stone from Pennsylvania. He was a newlywed. He had married his childhood sweetheart during his thirty days' leave before deployment to Vietnam. We later learned Stone's new wife was pregnant.

Like all Stingers, Stone volunteered for the gunships. It was the allure of being part of the most respected, elite, and dangerous platoon of any assault helicopter company.

A Wasp slick was nearby when the Stinger ship was shot down. They rescued the surviving crew. SP4 Stone's body was recovered. The downed, salvageable Stinger aircraft was still there.

A Chinook CH-47 tandem rotor helicopter was summoned to recover the disabled ship. Our mission that afternoon was to provide cover for the Chinook. If the enemy was still in the area, we would kill as many as possible.

Our pair of Stinger gunships approached the crash site. My ship, 470, was in the lead. As we got nearer to the downed ship, we surprised six Viet Cong who were scavenging the aircraft. When the VC heard us coming, they scattered.

I jumped on the intercom. "We got dinks running in different directions."

Our AC, Lt. Berman, responded, "Right Chief, I got 'em."

The AC called our wing ship, "Stinger 27, Stinger 42. Do you see the two dinks running to the south?"

Stinger 27 responded, "Roger that, we see them."

The wing ship broke off and started a rocket run on the two evading dinks.

Lt. Berman keyed the intercom. "Chief, I am going to use rockets on the four dinks running to the west. You and Owens back me up with the M-60s."

I responded, "Yes sir, copy that."

Our aircraft lined up on the four dinks from an altitude of about eighty feet AGL. The dinks were looking back at us with faces contorted in terror as they ran for their lives.

Owens and I leaned out of the ship into the airstream, firing our M-60s forward, past the pilots' doors.

The AC fired three salvos of the seventeen-pound rockets along the evading dinks' path. The rockets exploded right on target. The powerful explosions fragmented the dink bodies. Body parts spiraled in all directions.

The AC banked into a left-hand orbit around the dink remains.

I spotted one dink that had a leg and arm blown off, but was still alive. I opened fire and finished the job with hits to the body and head.

I sprayed the remaining bodies with machine gunfire. My rounds tore the partial bodies into shreds. Brains and guts spewed outward in a cloud of red mist.

We continued circling the crash site, inviting any dink to fire on us. We were ready. We wanted blood.

The Chinook arrived. The big ship settled to the ground, to allow the rigger to jump off so he could attach the lifting gear to the downed Stinger ship.

The huge Chinook was a tempting target for the enemy. The enemy could not resist.

The Chinook started taking fire from the tree line to the east. The Stingers reacted, rolling in on the tree line with rockets and M-60s blazing.

The brave Chinook rigger climbed up the crashed ship's fuselage and got on top of the mast so he could rig the lift line to the main rotor head. Enemy bullets began hitting all around him.

The Chinook door gunners retaliated with return fire.

The Chinook aircraft commander called our AC. "Stinger 42, Hook 33, we are taking intense enemy fire. Our ship is taking hits. We cannot stay and complete the mission if you guys can't take out the enemy, over."

Our AC responded, "Hook 33, Stinger 42, we are on the dinks. Just give us a couple of minutes, over."

"Stinger 42, Hook 33, roger that, two more mikes."

I spotted the enemy firing position and directed the Aircraft Commander to the spot. I began spraying the area with my lethal bullets.

The AC broke into left-hand orbits around the enemy position.

Owens moved to my side with the M-79 grenade launcher. As I continued firing on the enemy position, Owens fired the M-79 grenades.

The explosions rocked the jungle as the fragmenting shrapnel shredded enemy soldiers.

Our wing ship broke off and attacked the tree line surrounding the crash site. They neutralized the remaining enemy firing positions.

The rigger completed the job and signaled the Chinook to

ascend to a hover above the crash. The rigger attached the lift lines to the Chinook and climbed back aboard.

The Chinook carefully lifted the disabled Stinger warbird.

“Stinger 42, Hook 33, thanks to you and your crews for saving our butts, over.”

“Hook 33, Stinger 42, roger that. Thanks for taking our wounded bird home.”

“Stinger 42, Hook 33, anytime for you guys. Over and out.”

The Stinger guardian ships continued circling as the Chinook gained altitude and departed the area with the damaged Stinger ship, slightly swaying underneath.

Satisfied we had defeated the enemy and avenged SP4 Stone’s death, the Stinger warships departed the area.

More than a dozen dismembered enemy bodies littered the ground.

We left them to rot in the shimmering jungle heat.

CH-47 Chinook

CHAPTER 39 TURKEY SHOOT

An Americal unit engaged a large VC force near LZ East, twenty miles northwest of Chu Lai. That morning, my ship 470 and a wing ship were sent on a search and kill mission to the same area to kill off any stragglers.

It was unusual for the VC to form units. We usually encountered just a few VC and often just one dink.

The VC were not well trained, so we weren't sure what we would encounter. We arrived in the area and started patrolling. Almost immediately, I spotted a dink with a weapon running for cover. I keyed the intercom for the AC. "We got a dink on the left, bank left".

The AC responded. As the ship came out of the steep bank, we were taking fire. Green tracers streaked by my doorway.

The AC screamed, "Get that son-of-bitch."

I spotted the dink again as he was aiming his AK-47 straight at me. I returned fired. My M-60 rounds were right on target. The dink got hit multiple times as he flew backward from the bullet force.

Owens, the gunner, opened fire on the right side, targeting suspected hiding places.

Owens screamed, "Taking fire on the right."

The AC yelled back, "Owens, do you have the bastard."

In reply, Owens continued firing.

The AC banked to the right.

Owens stopped shooting.

He clicked on the intercom, "I got him". That was a rare event for the right-side gunner to get a kill.

The entire crew yelled congratulatory profanity while giving Owens the thumbs up. Owens was beaming like a teenager after getting laid for the first time.

The jubilation was short-lived.

We started taking fire again on the left. Both the AC and I spotted the dink at the same time. The AC banked hard left around the target. I started shooting a steady stream of M-60 fire. I followed my tracers as they hit the ground, closer and closer to the dink.

The bullet connecting links and spent cartridges were showering on the deck as my M-60 continued firing to kill the enemy.

I got hits at the center of the dink's body. He fell over, dead.

The wing ship spotted dinks and engaged. Shit, we were in a serious battle. Enemy bullets hit 470 in the tail boom.

Moments later, a bullet crashed through the co-pilot's chin bubble, barely missing the pilot.

The AC had enough of this crap. Both ships started strafing the entire area with rockets. The explosions stirred up the roach nest. Dinks started coming out of hiding places and running everywhere.

I screamed on the intercom, "I got multiple dinks. Taking fire on the...."

Owens jumped on the intercom and cut me off. "I got a bunch of dinks on the right."

It was clear we were in a major battle. We were taking fire, but more dinks were running than fighting.

The AC was communicating with the wing ship AC to get a quick plan.

Both ships flew to opposite sides of the battlefield to contain the dinks in the kill zone. We started left-hand orbits.

We moved closer to the center of the field of fire as we engaged each zone of enemy combatants.

The sky was full of tracers aimed at us. We took multiple hits.

We started a rocket run on several escaping dinks. The explosions were ripping dink bodies into pieces. The AC was getting lower to the ground as he maintained an optimal profile for aiming the rockets.

I saw a dink jump in a hole ahead of our flight path. I grabbed a grenade and pulled the pin. As we flew by the hole, I dropped the grenade.

Bullseye!! The grenade flew right into the hole.

I leaned out and watched behind us. The grenade exploded, and an arm came spinning out of the hole.

As we were gaining some altitude, a dink popped up in front of us and had dead aim on our ship.

The enemy's bullets riddled the bottom of the ship.

The Master Caution light illuminated in the cockpit.

We were losing hydraulic fluid.

The co-pilot switched over to the backup hydraulic system. The AC called our wing ship and advised of our status.

Our turkey shoot day was over. Our helicopters were not controllable without hydraulics. The backup system was designed to get us home.

For the first time, I patted 470 on the tail boom when we landed back at base.

I hoped Curtis McKinley would not mind.

Stinger 470

CHAPTER 40 CHANGE OF HEART

In early 1971, it was clear the United States would not achieve a decisive military victory over the NVA and VC.

The U.S. strategy shifted to the fortification of the ARVN's ability to take over war efforts against their enemies.

Meanwhile, the U.S. would continue to pursue an agreement with North Vietnam for the U.S. to "honorably" exit the war.

North Vietnam knew their victory was near. They just had to maintain pressure on U.S. forces at any cost.

Although the U.S. publically maintained the position that Vietnamization was working, the remaining U.S. soldiers in Vietnam knew it was a fantasy.

Those of us who were still fighting the war became burdened with a difficult paradox.

The elite aviation units were built upon unwavering patriotism to serve our country.

They imbued us with duty and honor to accomplish our assigned mission. It was our charter to protect and defend the highest principles of American military service.

In reality, we were in retreat. Moral support from our Country was decreasing. Military morale was plummeting. Leadership was in disarray. The widespread substance abuse within the military was endemic to that environment.

The dilemma begged the question, "What are we fighting for?"

I had done my duty. I had killed dozens of enemy combatants. I had taken part in destroying the enemy's ability to conduct war.

I needed to focus on one thing. Survive at all costs. There would be no glory coming home in a body bag.

From my perspective, morale within the Stinger enlisted ranks continued to deteriorate. The surviving Stinger veterans were rotating home. The few new guys were coming in with no knowledge or experience from the Stinger glory days.

McKinley's replacement as Platoon Sergeant was a veteran Stinger who was on his second or third tour. He had hundreds of personal kills.

Although the new Platoon Sergeant had the experience and credentials, it would take a messiah to resurrect the feeling of the Stinger glory days.

I believe that warriors who continue to fight for the sake of fighting, in the absence of a crusading cause, are admirable and incredibly courageous.

The reward is winning the ultimate game. Kill or be killed.

Then, there are the warriors who are inspired to fight for a worthy cause.

I had proved everything to myself I needed to prove. If the cause is righteous and makes sense, I would do it again.

That mindset did not make me a career Army man.

Sacrificing my life for nothing but pride or personal glory was not part of my thinking process.

CHAPTER 41 REFUSAL TO KILL

Americal troops encountered stiff enemy resistance from a village about thirty-five miles southwest of Chu Lai. They tasked a pair of Stinger gunships to assist the Americal unit in assaulting the village and rooting out the enemy.

My ship 470 and a wing ship were assigned the mission and scheduled to rendezvous with the Americal unit.

The grunts had flanked the village to prepare for a sweep from east to west. Before the sweep, our job was to orbit the village and take out any hooches that were camouflaged as firing positions. If any dinks attempted to escape outside the village to the west, north, or south, they instructed us to kill them.

We started the assault by attacking a large hooch suspected of harboring several dinks. The AC set up the rocket run. Owens and I stood outside the ship with one foot on top of the rocket pods for support so we could fire the M-60s in a forward field of fire. We began firing at the hooch as the AC started the rocket run.

Our wing ship continued orbiting the village, searching for escapees.

As our M-60 fire peppered the hooch, a flap in the hooch roof opened and a large-caliber machine gun started firing at us.

Taking fire!!

The enemy tracers streaked past us just a few feet away.

The AC increased the rocket launches. An enemy bullet hit our main rotor blade. The ship shuddered and began a vibration. The blade was whistling from the bullet hole as it rotated.

Our rockets hit with an enormous explosion. The AC broke off the run to the right. Our wing ship rolled in for a second rocket run.

Enemy machine gunfire had stopped.

As the wing ship was about to fire rockets, we started taking fire from multiple quadrants.

Both Stinger gunships sprayed M-60 fire on the entire village.

We broke off and exited the village area to check the aircraft's airworthiness with the damaged main rotor blade.

The rotor blades are constructed with a honeycomb design that isolates damage, which helps to keep structural integrity intact.

Our AC determined we were still airworthy.

The Americal commander started the ground sweep and requested the gunships continue watching for escaping dinks outside the village perimeter.

Several dinks started running everywhere. The gunship gunners killed multiple dinks as they attempted their escape. All the evading dinks had weapons.

Some dinks fired off rounds at us as they ran for their lives.

I heard and felt three or four hits in the center of our fuselage.

The grunts were in full engagement, moving from hooch to hooch, neutralizing enemy threats.

All escaping dinks were down.

We glided from body to body, peppering the corpses with gunfire to ensure they were dead.

Suddenly, a figure ran from behind a bush into the open.

The AC was yelling, "Get him, get him".

I was laying down fire, using my tracers to adjust my aim as the AC got lower and lower on the target.

We were nearly on top of the figure, presenting a narrow target profile. My bullets kept hitting all around him.

The AC screamed, "Goddamit, kill the son-of-bitch."

The figure was not armed; I did not see any military gear. As we got lower, the figure looked up, his face frozen with fear. He was crying. It was a young boy.

Something clicked in my mind. I could not pull the trigger. The months of training and lethal experiences shattered in a flash of humanity.

I radioed the AC. "This is just a kid. I will not kill him."

The AC did not respond. Owens looked away from me.

There was only silence.

The young boy ran away as fast as he could.

The operation was over.

The Americal commander released us and we returned to base.

No one spoke a single word to me about my refusal to kill the unarmed boy.

CHAPTER 42 CATHARSIS

Catharsis is defined as releasing, providing relief from strong or repressed emotions.

Something changed in my psyche when I refused to kill that young boy. I could kill so many times before. I knew unarmed women and children were probably killed during the many times I fired into hooches, jungle, and hiding holes in the ground.

I guess I dealt with those probable kills on a subconscious level.

The incident with the young boy was on a conscious level.

After months of witnessing the disintegration of military morale, widespread substance abuse, failing leadership, and the winding down of this spectacularly unpopular war; the refusal to kill an unarmed young boy running for his life was the turning point.

I decided. I was going to quit flying.

I spoke to the Stinger Platoon Sergeant and announced my decision. He told me I would have to see Captain Fisher (formerly Lt. Fisher).

I received word Captain Fisher would see me in the Officer's hooch.

I knocked on the door. Captain Fisher appeared. This time, he invited me in.

The Captain started, "I understand you want to quit flying."

I replied, "That is correct, Sir,"

The Captain responded, "So, you are quitting like you quit OCS."

I was silent for a few seconds. I was angry. But I needed to cool down and be rational.

I spoke up in a stern, determined manner, "Captain Fisher, I am grateful for the opportunity to be a Stinger. In some ways, it exceeded my expectations. I performed my duties to the best of my abilities. I worked hard so my crew could rely on me to ensure our aircraft was airworthy. I have thirty-six confirmed personal kills and many unconfirmed kills. My crew and I survived an inflight fire and subsequent crash. They awarded me the Purple Heart, Air Medal for Valor, Army Commendation Medal, and the Air Medal with seventeen Oak Leaf Clusters. Our Commander-in-Chief has ordered the winding down of the War. There is no longer any purpose in continuing to kill except for defense. It is not for patriotism. It is not for the American way of life. I choose not to continue risking my life in a gunship for no purpose or cause."

I summed it up, "Sir, I have done my duty. I respectfully request to be relieved from flight status."

Captain Fisher was silent. He looked at me for about a minute.

He then responded, "Your request is approved."

I arose, saluted the Captain, and started for the door.

The Captain added, "Allen... that was a good catch on that Jesus Nut."

We both smiled.

I said, "Thank you, Sir."

I never saw Captain Fisher again. He rotated back to the States a few weeks later.

CHAPTER 43 REASSIGNMENT

As an E-5 NCO, I had some latitude on how to spend the remaining time of my tour in the war zone. Because of the overall lack of leadership focus and general melee of the war winding down, I could manipulate the best situation for myself.

I was a little surprised most people in the unit were aware of my history as Crew Chief in the Stingers. So, they showed me some unexpected respect.

We had a new First Sergeant, whom I had not met. I squared up my appearance and went to the Orderly Room to discuss a reassignment in the unit.

I did not want to go to the maintenance platoon. A spot in the Orderly Room was my aim.

Top turned out to be a pretty reasonable fellow.

He was aware and impressed with the 523 crash and burn story. He said he would talk to the CO and see what could be done.

I requested permission to move into the NCO hooch by the ocean.

Top consented.

The NCO hooch had individual rooms. Although primitive, the newfound privacy was welcomed.

While I waited for the CO's decision for my new assignment, I did some remodeling. The 2.75 rockets were packed in knotted pinewood crates. The crates were easily dismantled, producing some very nice 1/2 inch thick paneling.

I dismantled several crates, scavenged a hammer, saw, and nails, and went to work. In a few hours, I had a pretty decent living space.

I bought a little black-and-white TV to receive our only available channel; Armed Forces Network. I also picked up an oscillating fan at the PX.

I built a couple of shelves and an elevated support system to raise my bunk, providing more storage room underneath.

I could step out of the hooch, right onto the beach. With the increased time on my hands, I could work on my body surfing technique when the JP4 fuel contamination was not too bad.

Offshore storms would sometimes send in eight to twelve-foot waves.

As protection against any sappers sneaking in and slitting my throat in the night, I installed interior locks on the door. Of course, my weapon was always locked and loaded by my bunk.

The NCO hooch was right next to the JP4 tank farm. If sappers blew up the closest tank that would probably be the end of the war for me.

There were other hazards I couldn't even imagine.

Top got back to me in a few days.

The CO approved my reassignment to the Orderly Room.

With my status change, logic would suggest I had a pretty good chance of getting out of the war alive and in one piece.

I was still in a war zone. Anything could happen.

CHAPTER 44 SERGEANT OF THE GUARD

Orderly Room duty was fairly laid back. We prepared a lot of Americal obscure reports for various purposes.

It was a very different perspective than flying in gunships. We would get updates from operations regarding the flight status of the ships and crews. I liked still being a part of that.

I witnessed an Americal full bird colonel stomp into the Orderly Room one day demanding to see our CO. The Major came out of his office and the Colonel tore into him. I had never seen a buck private get chewed out as bad as our Major did that day. It was symptomatic of what was happening on a macroscale.

The war was winding down in a lost cause. Tensions were high. Careers were being delayed or destroyed. Aircraft parts and everything else was in short supply. Most guys did not want to be there.

As an NCO, they charged me with a few supervisory duties.

The most memorable was Sergeant of the Guard.

Americal assigned me a jeep and a section of the guard towers on the base perimeter for a few night watches.

I would take an occasional drive up and down my assigned stretch of the perimeter, checking on each guard tower.

Almost all the guys manning the towers were infantry.

I was shocked to learn drug use during guard duty was routine, particularly heroin.

The greatest risk in that duty was not an enemy attack. The risk was managing over twenty battle-worn, heavily armed druggies in the middle of the night.

The last thing I wanted to do was to piss off a drugged-out grunt who had a loaded automatic weapon in his hands.

I decided my best course for survival would be not to take an authoritative approach with these guys. I would just check on each tower occasionally and have a conversation.

"Hey guys, just checking to make sure at least one of you are awake."

The guys were eating C-rations and throwing the cans or pouches on the floor of the tower. The ten-foot guard towers had a built-up dirt berm at the front. That provided easy access for the world's largest rats. The rats could easily be mistaken for huge cats. The rats would scavenge the remains of C-ration food on the floor.

So, among the enormous rats, heroin needles, and the constant smell of killer pot, the nights wore on in the base perimeter guard towers of Chu Lai, Vietnam.

After my first night, I decided I needed to report the drug use problem through my chain of command.

Americal had not given me the chain of command for the guard detail. I did not know who my immediate superior was.

After considerable effort, I got the contact phone number for the Americal Duty Officer. The sergeant for the Duty Officer told me they would get back to me. They never did.

I discussed the drug use problem in the guard towers with our First Sergeant. Top just blew it off. He offered no solutions. He said he would not bring the issue to the CO's attention. Top insisted Americal was the chain of command.

Everything was in disarray. Everyone was just trying to stay low and get out of the war alive. Considering the totality of the situation, I decided I would do the same. I would not confront the grunt addicts in the guard towers.

With a couple of months left on my tour, being a "white knight" in the darkness was not a viable option. My primary objective was to survive the Vietnam War.

One night, when I was Sergeant of the Guard, I got a little crazy, probably from second-hand reefer smoke.

Some dink was riding a bicycle up and down the road outside the perimeter in the middle of the night. He looked like a VC probing the perimeter to me. I fired an entire magazine of M-16 rounds at the dink. The bullets hit all around him. I didn't want to kill the dink, which could have repercussions against me. There were no standing orders dictating rules of engagement.

We shot off a few flares to brighten things up. That's when I saw a gigantic rat looking straight up at me in front of the tower. I loaded another magazine and plugged the rat with one shot.

The tower was only a few hundred feet away from the main gate, and the Military Police building. All the racket caused a couple of MPs to come out and look. They were probably high, too. These guys did not draw weapons or act like anything was wrong. They were just curious. I guessed they were used to the commotion at night or they just didn't care.

CHAPTER 45 FRIENDLY FRAGGING

The guy in the room next to mine in the NCO hooch was a Sergeant First Class. He was a supervisor at the aircraft maintenance hangar.

This SFC was relatively new. They brought him in to get the substance abuse by maintenance workers under control.

The Sergeant implemented various measures to slow down the supply of drugs. That did not make this guy popular.

A few entrepreneurial pushers had set up drug distribution operations. They did not want their businesses to be disrupted. Of course, everyone was armed with automatic weapons and well trained on how to use them.

One channel of drug supply was the grunts. All the grunts had access to fragmentary grenades.

A rocket attack hit us one night. Several rockets were hitting the tank farm and various spots in our company area. We did not have any huge explosion fires from the JP4 tanks, but other rockets hit close.

It rattled me rather badly. I stayed pretty jumpy for a while. I was suffering from the short-timer blues. Rising hope of surviving with rising fear of being killed.

A few nights later, a loud explosion detonated very near our hooch. I was lying asleep in my bed. The concussion knocked me out of bed onto the floor. Everything on my wall shelves was knocked down and lying on the floor.

Clad only in boxers and a t-shirt, I crawled out of the hooch to the sand and wiggled into the bunker next to the hooch. It scraped me up pretty well. There were no other explosions.

After several minutes, everyone started coming out of the bunkers. Someone said, "That was really close".

One guy grabbed a flashlight and everyone started looking for an explosion crater in the sand.

Finally, someone yelled, "it hit here".

The crater was under the maintenance supervisor's room in my hooch. My bed was just a few feet away from the blast. The Sergeant was not there. The explosion was from a concussion grenade. There was no damage to the hooch.

I was very lucky. A fragmentary grenade would most likely have killed me or left me severely wounded.

Someone wanted to send a message to the maintenance supervisor.

No one challenged the assumption the fragger was after the maintenance supervisor. It seemed to be common knowledge. The Sergeant was a marked man.

I got back into my room and picked up all the stuff on the floor.

The next day, I scavenged three planks of PSP to install on the floor and walls of my room. PSP or perforated steel planking were made of concrete and steel and used for aircraft runways.

If there was another fragging, it might be a lethal fragmentary grenade. I hoped the PSP planks would stop the exploding shrapnel from killing me.

I was determined to survive the Vietnam War.

A slick Aircraft Commander, Captain Ron Colby, was assigned to investigate the fragging incident. I knew Cpt. Colby from various flight operations. The Captain would not discuss the drug trade implications.

I heard the maintenance Sergeant was defiant about the first scare attempt. He reportedly increased pressure on the maintenance druggies.

This is speculation on my part. The drug epidemic seemed to be comprehensive and coordinated. The supply was plentiful. The price was cheap based on what I was hearing.

After discovering the heroin convoy via hooch maid several weeks earlier, it made some sense. The enemy could easily get drugs into military bases in Vietnam. U.S. troops were disgruntled and demoralized. The drugs were plentiful, cheap, and highly addictive. The enemy could negatively affect the U.S. fighting force without losing a single NVA or VC soldier and without firing a shot.

CHAPTER 46 NO PLACE WAS SAFE

We had a vicious rocket attack one day with over twenty rockets hitting near the flight line. I heard later they hit the medical center.

I got a ride and went to the med center to check on Julie, the beautiful nurse I had met when we visited Curtis McKinley on his deathbed.

It saddened me to find the med center demolished. The huge Red Cross that was painted on the roof of the building under the protection afforded by the Geneva Convention was barely discernible lying on the ground among the ruins.

The Associated Press reported:

Rocket Wrecks U.S. Med Post - Saigon (AP) *The medical dispensary was smashed during a 20-round rocket barrage on the Chu Lai base camp, headquarters of the U.S. Americal Division. Field reports said one of the 100-pound missiles tore through the roof of the first aid station about 4 p.m. The reports said there were no patients in the dispensary, but the entire American staff was killed. The U.S. Command's security regulations prohibit disclosing specific details about casualties.*

I could never confirm if Julie was among the dead.

CHAPTER 47 CHARGE OF QUARTERS

At night, they assign an NCO to be CQ or "Charge of Quarters". The duties included monitoring all unit areas for suspicious activities, security, the well-being of the troops, and protection of unit assets. The CQ reported to the Duty Officer in Charge.

The NCO wears an armband identifying him as CQ and is assigned a jeep. I pulled CQ duty a few times when I was a temporary E-5 as a Class Leader in AIT. So, I knew the drill.

The second time I pulled CQ duty at the 116th AHC, I received a strange report at about 2300 hours. Personnel were forming long lines at the supply depot near the flight line. I was not aware of any combat operations that night. I jumped in my jeep and drove to the supply depot to investigate.

Sure enough, guys were standing in line waiting to get inside of the building. When I got out of the jeep, the guys saw my armband and I could hear, "Oh shit, it's the CQ".

I asked the guys in the line what was going on. Everyone clammed up.

I had my 45 caliber pistol strapped on in the event of any threats. I walked into the building. The line ran up the stairs to the second level. I heard shuffling behind me. I turned to see guys scattering, briskly evacuating the area.

That put me on a higher alert.

I walked up the stairs, brushing by the guys in line. Near the top of the stairs, I peered into the open bay.

In the center of the bay were five Vietnamese women lying naked on floor pads while soldiers were on top having intercourse. As the soldiers finished, the women would get up, walk over to one of a few pails of water and clean themselves.

They returned to their floor pads and laid down to prepare for the next customer.

I spotted a supply sergeant who outranked me. As CQ, I was authorized to take action.

I walked over to the sergeant and asked what was going on. The sergeant knew me from the Stingers, so he was forthright. He explained they had a little enterprise going. No one was getting hurt. No "customers" had access to supply property. He said he would take responsibility for the safety and security of the building.

It was a stretch. If I ignored the event, I was putting my neck on the line along with the supply sergeant. I thought about it for a couple of minutes. I rationalized this was probably the only thing these guys could be happy about for a little while. They could get killed by a rocket attack tomorrow. I let it go.

I told the supply sergeant he needed to wrap it up by midnight. He agreed.

In an environment that is rife with degradation, death, suffering, loss of hope, fear, and anger; moral values seem to dissipate into nothingness.

March 1, 1971–*In protest of the Vietnam War, a radical group called the Weathermen, plants a bomb in the Capitol building in Washington, D.C., causing $300,000 in damage, but no casualties.*

CHAPTER 48 SECOND FRAGGING

Just a couple of weeks after the first fragging, I returned to my hooch one day and found a note on my door. It read, "Allen–get out now".

I did not know what that was all about. With all the craziness that was going on, I just dismissed it.

Two days later, at about 2100 hours, a fragmentary grenade was thrown at our hooch.

This time, the grenade hit an electrical line that was strung between the hooches. The grenade fell next to a pilot's hooch. There was a group of slick pilots playing cards around a small card table inside the hooch.

The grenade exploded.

Both the maintenance sergeant and I were in our rooms. Other NCOs were also in their rooms. The explosion prompted all occupants to get low and run to the bunker. I nearly ran into the maintenance sergeant as he bolted past my door for the exit.

There was a lot of commotion around the pilot's hooch about thirty feet from the NCO hooch.

Someone started yelling for a medic.

One of the slick pilots was hit in the eye by shrapnel. The remains of the punctured eye hung out of the eye socket.

The young pilot's war and flying career were over.

As they administered first aid to the stricken pilot, Captain Colby came charging on to the scene. The Captain was livid. He seized anyone who was near and viciously interrogated as if everyone was a suspect.

I will never forget the Captain looking at me with intense vengeance flashing in his eyes. After recognizing me, he moved on to another suspect.

They made no arrests.

I heard they theorized that someone hired a grunt from across the road to kill the maintenance sergeant.

Two days later, they transferred the maintenance sergeant out of the 116th AHC.

Captain Colby met his fate just a few weeks later.

The Captain was flying a slick as AC. Intense enemy fire hit the slick on a mission.

Captain Colby crash-landed. The hard landing bent the landing skids. The crew escaped the damaged craft.

The main rotor sweep was perilously low to the ground because of the damaged skids.

Captain Colby ran away from the ship, staying low below the rotor sweep.

The Captain stood up too soon.

The end of the main rotor blade hit him in the abdomen with immense force.

The Captain was nearly decapitated. His body was thrown forty feet away.

He probably never knew what hit him.

CHAPTER 49 LAM SON 719

The American withdrawal from the Vietnam War had picked up speed. The ARVNs needed to assume more strategic responsibility in the defense of their country.

The NVA and VC supply line, known as the Ho Chi Minh trail, ran through the eastern border of Laos adjoining South Vietnam. The strategic thoroughfare had to be cut off.

Any offensive planning by the U.S. was limited by the passage of the Cooper-Church Amendment on December 29, 1970, which prohibited U.S. ground forces and advisors from entering Laos. The U.S. however, could provide the ARVNs logistical, aerial, and artillery support in Laos.

U.S. Army General Creighton Abrams Jr. assumed command of U.S. Military Assistance Command-Vietnam (MACV) in July 1968, succeeding General Westmoreland. General Abrams maintained a positive belief the South Vietnamese military would prevail after the U.S. had exited the war.

The U.S. and South Vietnamese High Command conceived Operation Lam Son 719 to be an ARVN operation to destroy key points of the Ho Chi Minh trail and a major supply depot in Laos.

U.S. Joint Chiefs of Staff Chairman, Admiral Thomas H. Moorer, maintained a rosy outlook for the ARVN's success in the massive action.

Army Chief of Staff, General William C. Westmoreland was less optimistic.

President Nixon and Assistant to the President for National Security Affairs, Dr. Henry Kissinger saw Lam Son 719 as an opportunity to strike a decisive blow against Hanoi.

Both the President and Dr. Kissinger were encouraged by Admiral Moorer and General Abram's positive outlook for success.

In contrast to General Abrams' sunny outlook, South Vietnam President Nguyen Van Thieu had shown indecisiveness in previous operations.

President Thieu was reluctant to commit most ARVN forces to Lam Son 719.

South Vietnamese Lt. Gen. Hoang Xuan Lam was designated the overall commander of Lam Son 719. The general, a firm ally of President Thieu, was in charge of South Vietnam's five northern provinces, the military region designated as I Corps. But Lam was primarily an administrator, not a tactical corps commander. The general hesitated on major decisions.

Lam Son 719 was designed as a complex battle plan requiring exceptional leadership.

The U.S. Army Vice Chief of Staff, General Bruce Palmer, remarked that "only a Patton or McArthur could be successful with such an intricate plan."

General Lam and his staff were unprepared for the challenges of directing multiple divisions and orchestrating combined arms operations.

The Laos operation would be conventional warfare, not the smaller insurgent engagements with Viet Cong guerrillas that South Vietnam's leaders were used to overseeing.

There is speculation that ARVN forces were riddled with corruption. Some researchers claim there were 100,000 "flower soldiers" listed on ARVN personnel rosters. These were South Vietnamese civilians who paid ARVN commanders not to serve in the military.

In 1994, an audit of the Iraq army uncovered 50,000 "ghost" soldiers on its payroll. Historically, it might be inferred the ARVN personnel allegation is plausible.

In other instances, dead soldiers were kept on rosters so ARVN commanders could collect the dead soldiers' salaries.

Nine ARVN divisions were available, yet withheld from the most crucial battle of the Vietnam War.

General Lam commanded 30,000 ARVN troops. Only 17,000 were made available for Lam Son 719.

There is speculation the forces Theiu and Lam refused to commit to Lam Son 719 were not combat-ready.

South Vietnam's best units the Airborne Division, Marine Division, and 1st Infantry Division — were part of Lam's force. The incursion force was assembled near the Khe Sanh Combat Base in South Vietnam. Their objective was Tchepone, Laos, a logistical staging point along the Ho Chi Minh Trail, about 25 miles from the Vietnamese border.

Trucks from North Vietnam carrying war material would be unloaded and repacked on smaller modes of transportation at Tchepone.

The southern portions of the Ho Chi Minh trail became a series of pathways and primitive dirt roads. The trails were impassable for larger motorized vehicles. Bicycles, pack animals, and individual porters were used to carry the loads on the rest of the trip into lower South Vietnam.

In late January 1971, U.S. Army assault helicopter companies with support units began moving north to Quang Tri. The intent was to form a massive armada of aircraft and men to support ARVN troops for the Laotian incursion.

Lam Son 719 started February 8, 1971.

It was imperative General Lam demand speed and aggressiveness from his commanders. However, the pace was slowed by South Vietnamese command-and-control problems.

Instead of being at the center of operations, General Lam

established his forward headquarters in Dong Ha, 30 miles from most of his subordinate units' command posts at Khe Sanh.

The distance required most communication be conducted by telephone and radio instead of face-to-face.

Poor coordination and lack of clarity in orders prompted confusion and hesitation.

The ARVN Airborne Division, which had never operated in a formation larger than a brigade task force, experienced similar issues.

Despite being a former military general, President Theiu did not have the needed military tactical experience. In denial of his lack of tactical expertise, the President assumed the role of a tactical leader.

On February 19, Thieu instructed his commanders "to conduct search operations near their present positions."

This guidance created an overabundance of caution. As a result, many ARVN units stayed on their firebases and made little progress.

The White House and American military advisors repeatedly attempted to advise President Thieu to correct his errant ways.

No amount of pressure from the U.S. advisory chain could overcome Thieu's arrogance.

When it became clear General Lam was not capable of managing the massive Lam Son 719 operation, another ARVN general was called in to replace Lam.

General Tri was a competent commander who led the ARVNs in the successful Cambodian operation in 1970.

General Tri was killed in a mysterious helicopter crash while flying to I Corps to assume command of Lam Son 719.

The actual cause of the crash was never investigated.

The massive communist infiltration of the South Vietnamese military and government would have given the enemy ample opportunity to sabotage General Tri's helicopter.

General Tri's untimely death was instrumental in the eventual defeat of South Vietnam.

General Lam's continued incompetence hastened the Communist victory in Lam Son 719.

The South Vietnamese lack of tactical prowess and poor organization, despite U.S. best efforts to advise, gave the numerically superior NVA the advantage.

After the war, the North Vietnamese claimed they had amassed some 60,000 troops along Route 9 when the ARVNs crossed the Laotian border. They included five main force divisions, two separate regiments, eight artillery regiments, three engineer regiments, three tank battalions, six anti-aircraft regiments, eight sapper battalions, with logistic and transportation units.

According to North Vietnamese historians, the Lam Son 719 battle was described as "our army's greatest concentration of arms and forces, up to that point."

The North Vietnamese had amassed this combat power for a larger purpose than simply defending their critical supply route. They were determined to seize an opportunity to fight a decisive battle on advantageous terms, destroy a large portion of Saigon's army, and thoroughly disrupt and discredit Vietnamization.

Showing the importance of the campaign, Hanoi sent Colonel-General Van Tien Dung, Deputy Chairman of the Politburo's Central Military Party Committee and Chief of the General Staff, to the front to oversee operations.

Previously, on January 31, in an address to the North Vietnamese troops on the Route 9-Southern Laos front, the Communist Party Central Committee made clear the operation's importance and objectives.

The following message was broadcast to all North Vietnamese units.

"The coming engagement will be a strategically decisive battle. We will fight not only to retain control of the strategic transportation corridor but also to annihilate a number of units of the enemy's strategic reserve forces, to deal a significant defeat to a portion of their "Vietnamization" plot, to advance our resistance effort to liberate South Vietnam and defend North Vietnam, to gloriously fulfilling our international duty, and to hone our main force troops in the fires of combat. Our Army must certainly win this battle."

The North Vietnamese command was committed to an all-out effort to counter the ARVN incursion of Laos, supported by U.S. helicopter slicks and gunships.

When the ARVNs executed their ill-fated battle plan of maintaining and defending stationary firebases, the NVA seized the moment and immediately went on the offensive.

North Vietnamese units attacked ARVN firebases and pounded them with artillery fire. The firebases became surrounded by NVA anti-aircraft weapons that restricted U.S. helicopter and close-air support.

On March 2, 1971, the first U.S. assault helicopter companies, composed of forty helicopters, were briefed on the mission to assault LZ Lolo in Laos.

They were advised that B-52s were currently conducting bombing runs on the area and the LZ would be secure. For that reason, no gunships would provide cover for the slicks inserting ARVN troops.

One briefer, who was a U.S. senior officer, said the LZ insertion would be a "piece of cake"

The area was crawling with NVA. After the B-52 bombing stopped, the NVA worked through the night and early morning, preparing the area for a massive ambush.

On March 3rd, at 1000 hours, the LZ insertion began. The 71st AHC "Rattlers", would be the first to enter the LZ area.

The first slick started taking intense enemy fire about 150 yards away from touchdown in the LZ. Under heavy fire, the ARVN troops were offloaded and the first ship struggled to escape the area as quickly as possible.

After departure, the AC discovered the cyclic control was jammed and could not control airspeed. They headed for Khe Sanh to make an emergency running landing.

The Khe Sanh air traffic controller advised the stricken ship they would be "Emergency 3" for landing. Two other battle-damaged ships were inbound from other Laotian missions. The first ship was on fire. The second ship had its tail rotor shot out.

The Rattler hit the ground at 90 knots and skid to a safe stop.

It was discovered a piece of shrapnel had lodged in the cyclic bell crank.

The ship was repaired and the Rattler crew continued flying missions the rest of the day.

The airmen flying into the deadly ambush quickly realized the mission briefing from the previous night was bogus.

The second ship, going into LZ Lolo, started taking enemy fire at 100 feet. The windshield exploded, and the AC took three hits in the center of his chest chicken plate, and three more in his armored seat. The crew chief was hit in the leg. Then the tail rotor got hit, and the aircraft began to spin. The aircraft continued to take hits, destroying the hydraulics system. With both pilots on the controls, they could control the aircraft.

For a moment, the crew thought they might survive. Seconds later, the engine failed, and then the transmission seized. The Rattler slick crashed into the trees.

The ARVNs and aircraft crew escaped the crash. The crew took refuge behind a log. The NVA were all around the area, searching for the survivors. The crew only had one 38 caliber pistol for defense.

The ARVNs secured LZ Lolo. More American ships and Dust-off reached the LZ. The crew from the first crashed ship was evacuated.

Another slick approached for a landing and was hit by enemy bullets. The ship caught on fire and crash-landed in the LZ. The landing skids were damaged, causing the main rotor blades to sweep at a low level.

The co-pilot exited the aircraft and ran under the main rotor sweep. The co-pilot raised up too soon. The main rotor blade hit the co-pilot in the head, killing him.

Later that morning, the 174th AHC Dolphin slicks were tasked to the LZ Lolo mission as replacements for all the previous losses earlier that day.

Dolphin aircraft 639 was inbound and called for smoke to identify the LZ. They saw the smoke and approached the area. Dolphin 639 was immediately shot down in flames.

The NVA was monitoring the ARVN radio frequencies and would pop smoke in their locations to lure unsuspecting crews to their deaths.

The co-pilot of Dolphin 639 survived the crash and escaped, evading the enemy. He walked several kilometers back to the South Vietnam border.

The door gunner was killed in the crash. The crew chief survived and stayed with the injured aircraft commander.

The 174th AHC Sharks gunship platoon engaged the enemy to support the LZ LoLo mission on March 3rd and 5th.

A Shark crew chief was hit by a .51 caliber bullet. After the massive projectile burst through the ship deck, it struck the crew chief's lower back and exited at his shoulder.

Undeterred by the gaping wounds caused by the large caliber bullet, the crew chief continued to fight. After killing several more NVA, the crew chief was hit again by an AK-47. The Shark crew chief survived and lived to tell the story.

After the events of March 3rd and 5th, only one Shark gunship remained. The surviving Shark was later attached to the Stingers until they could rebuild the Shark's unit.

Eleven combat helicopters were shot down at LZ Lolo and thirty-five were damaged.

Some of the most intense anti-aircraft fire of the Vietnam War was experienced during Lam Son 719.

The 116th and 176th AHC companies were tasked to supplement units at Quang Tri to make up for battlefield losses.

On March 4th, 1971, with brief notice, the entire Hornet force was packed and moved out for Quang Tri, just fifteen miles south of North Vietnam.

The Yellow Jackets, Wasps, and Stingers flight platoons took off in the early morning on the way to the enemy's doorstep.

The maintenance and other support platoons would convoy on Highway 1, through Da Nang and the ancient city of Hue.

I was riding in the back of an open deuce and half-ton truck with a bunch of new guys and a mountain of supplies. We were armed with M-60s and M-16s.

The new guys were oblivious to the danger of riding north on the exposed primitive highway. As we entered a canyon with high rock walls on both sides, I hunkered down into a suitable firing position and readied my M-60 and ammo supply.

The new guys were dumbstruck, oblivious to any threat. I instructed them on the possibility of an enemy ambush. They soon caught on and followed my lead.

We rolled into Quang Tri late in the day. Everything was in chaos. They called the Hornet flight platoons out on a mission when they arrived that morning.

The ARVNs were encountering heavy resistance. U.S. aircraft were receiving intense enemy fire. The enemy was pulling out all the stops to repel the ARVN force.

As the days wore on, the situation became much worse. The NVA stepped up rocket attacks on the Quang Tri base camp. We had to eat, sleep and work within a couple of minutes' dead run to a bunker.

The enemy zeroed in on the 116th AHC area. The Russian-made 122mm rockets started coming in before midnight.

The first rocket hit a hooch with eleven guys sleeping inside.

It was very near my assigned hooch. The concussion knocked me out of bed. All I could do was grab my M-16 and run wearing one flip-flop, T-shirt, and boxer shorts. I dived into the bunker, which was already at capacity.

There was no room for me to get deeper into the limited protection of the bunker. I was exposed as I kneeled at the opening.

Another rocket launched. Someone said, "Here comes another one." I heard a soldier sobbing. Another soldier started praying out loud.

The rocket hit next to my hooch. The blast was so close, my hearing was temporarily impaired. I was dazed.

Pieces of shrapnel and debris pelted the bunker.

I could barely hear a soldier running by, screaming, "They are coming through the wires".

We were being overrun.

A Cobra gunship, parked nearby, initiated a rapid start-up. It lifted off with no lights. I could hear the ship heading towards the perimeter in the pitch-black darkness to find and repel the enemy.

I refused to die while kneeling in a bunker.

With one flip-flop and one barefoot, my M-16 with one magazine, I crawled out of the bunker entrance and ran to the perimeter to make my stand.

I assumed a firing position in a small ravine about twenty feet from the set of barbed wire fences that comprised the perimeter. I only had around fifteen rounds of ammo.

I got lucky.

The enemy attacked an area of the perimeter about 100 yards from me.

The guys in the guard towers killed the NVA as they cut through the perimeter wires. The attack was over.

The experience numbed everyone that survived.

I rummaged through the debris that was my hooch. I retrieved a blanket, ammo, my fatigues, and boots. I dressed and found a spot free of rats and tried to get some sleep.

Sleep did not come.

In the early morning's first light, I began searching for food and water.

The calm, humid air reeked of the smell of human death.

I saw a warrant officer searching in the rubble of the first hooch that had housed eleven sleeping soldiers before taking a direct hit.

The officer was dragging a duffel bag. He picked up a splintered board from under an overturned bed. The board had a long section of large intestine entwined around it. He carefully unraveled the intestine and placed it in the duffel bag.

I watched as he retrieved more body parts and placed them in his duffel bag.

My mental and physical state at that moment was at the lowest point. The lack of food, water, and sleep weakened me.

The rocket attack shook me to the core. I could still hear the frightened soldier praying as the lethal rocket was racing inward.

The searing memory of those eleven dismembered soldiers in the early morning light scarred my soul for eternity.

I just had to hang on and find a way to survive.

The NVA hit the 116th area three more times, both day and night.

The Quang Tri attacks left me shell-shocked.

For many years, after the war, I would duck for cover any time there was a loud bang nearby.

CHAPTER 50 THE ARVN RETREAT

The plan called for the ARVNs to advance deeper into Laos and capture the supply depot at Tchepone. The firebases would provide artillery support for the ARVNs as they advanced.

Before the launch of Lam Son 719, it was discovered North Vietnamese intelligence had stolen complete ARVN plans for the invasion. The NVA were prepared.

As U.S. helicopters and fixed-wing aircraft crossed into Laos, they encountered heavy air artillery up to 10,000 feet.

For the first time, outside of North Vietnam, the communists used Russian-made shoulder-fired missiles and deadly SAM missiles.

The ARVNs reached Tchepone and destroyed some enemy munitions and supplies. They failed to hold Tchepone. The NVA forced the ARVNs to withdraw in just a couple of days.

ARVN commanders believed their mission had been "completed" and advocated for withdrawal.

Even as the NVA gained the upper hand, General Abrams continued to send positive reports to Washington. He believed General Lam and others would get their act together and become more aggressive. It was a major miscalculation.

Conditions on the ground deteriorated further on March 9, when the South Vietnamese president unilaterally ordered a withdrawal of all ARVN forces.

Senior U.S. leaders had hoped the ARVNs would remain in Laos through April.

The ARVNs lost 7,500 soldiers out of 17,000 committed to the fighting. The surreal casualty rate was 44%.

The ARVNs had been trained by top-notch U.S. military advisors.

The American people financed the South Vietnamese military and government for ten years.

Even with all the support and money, the South Vietnamese leadership squandered their country's future.

Lam Son 719 was a cataclysmic failure for the South Vietnamese.

After the United States had sacrificed over 55,000 American lives and spent nearly one trillion 2021 dollars to support South Vietnam's independence, President Thieu lost his nerve and capitulated. He cited ARVN casualties and disorganization in the Lam Son 719 operation as the reasons for the retreat.

By March 14, the NVA outnumbered ARVNs to 2 to 1. The superior NVA force attacked with renewed vengeance.

Retreating ARVNs turned into a rout.

ARVN soldiers were removing all military gear and uniforms to pass as civilians.

The ARVN command structure disintegrated.

Frightened ARVN soldiers were running for their lives as the NVA retaliated with terrifying ferocity.

U.S. Helicopter crews were tasked to evacuate the fleeing South Vietnamese while taking intense enemy fire.

There were 107 U.S. helicopters shot down and an additional 618 damaged.

The slick crew chiefs and gunners had to kick desperate ARVNS off landing skids so the overloaded helicopters could fly in the high-density altitude.

It got so bad, the chopper crews would coat the landing skids with grease and oil so the retreating ARVNs could not hang on.

The ARVNs abandoned 1,500 radios, 57 tanks, 87 armored personnel carriers, and over 500 crew-served weapons.

The final heartbreak–U.S. casualties were 253 dead or missing and 1,149 wounded.

Most of the casualties were helicopter crews.

March 23, 1971 - *National Security Adviser Henry Kissinger admitted to President Richard Nixon that Lam Son 719 "comes out as clearly not a success." The failure of Lam Son 719 was called by one scholar "the military turning point of the war."*

April 1971 - *The Vancouver Indo-Chinese Women's Conference (VICWC), conducted a six-day protest against the Vietnam War, gathering close to a thousand women.*

April 19–23, 1971 - *Vietnam Veterans Against the War (VVAW) stages operation Dewey Canyon III. 1,000 camping on the National Mall in protest.*

April 22–28, 1971–*VVAW (and John Kerry) testify before various congressional panels.*

April 23, 1971 - *Members of VVAW threw away over 700 medals on the west steps of the Capitol building in Washington to protest the war. The next day, antiwar organizers claimed that 500,000 marched,* making this the largest demonstration since the November 1969 march.

April 24, 1971 - *Peaceful Vietnam War "Out Now" rally on the National Mall, Washington, D.C., with 200,000-500,000 calling for an end to the Vietnam War, 156,000 participate in the largest demonstration so far on the West Coast, in San Francisco.*

April 26, 1971 - *More militant attempts in Washington, D. C. to shut down the government are futile against 5,000 police and 12,000 troops.*

May 3, 1971 - *15,000 soldiers and police arrested more than 7,000 persons protesting the war in Washington.*

May 5, 1971 - *1,146 protesters against the war were arrested on the U.S. Capitol grounds trying to shut down the U.S. Congress. This brought the total arrested during the 1971 May Day Protests to over 12,000.*

May 13, 1971 - *The Paris Peace Talks between North Vietnam,* South Vietnam, the Viet Cong, and the United States enter their fourth year. Little or no progress had been made.

CHAPTER 51 THE 116TH RETURNS TO CHU LAI

The 116th Hornets were ordered to return to Chu Lai. It was good to get out of Quang Tri.

My original rotation date was getting closer. I was short.

Guys started getting early rotation dates. My hopes were high, I would go home early. I could smell and taste that steak dinner, returning vets received as a token of appreciation for service.

The accumulation of the Quang Tri shelling, the friendly fragging incidents, all of my Stinger experiences, and frequent incoming rocket attacks had taken a toll on me. I knew something was wrong, but I could not quantify the problem.

I recognized I had lost my sense of humor. I suspected everything and everybody. I felt morally bankrupt.

That eighteen-year-old soldier that arrived in Vietnam nearly a year earlier, full of excitement to fly with the heralded gunships, no longer existed.

I just had one encompassing focus. I had to survive and escape Vietnam.

As my original rotation date drew near, my acute anxiety increased.

The time got shorter. Still no early rotation date.

Finally, I got my orders. They ordered me to depart three days short of a full year in Vietnam.

There were no farewell parties or getting drunk with buddies. The old-timers were already gone or just traumatized.

I just needed to stay alive a few more days.

June 13, 1971 - *The New York Times began publishing the Pentagon Papers. A 1996 article in The New York Times said the Pentagon Papers had shown the Johnson Administration had "systematically lied, not only to the public but also to Congress" about the United States' political and military involvement in the Vietnam War.*

June 24, 1971 - *The Mansfield Amendment, authored by Senator Mike Mansfield, was adopted by Congress. The amendment urged withdrawing American troops from South Vietnam at "the earliest practical date"—the first time in U.S. history that Congress had called for the end of a war.*

CHAPTER 52 FREEDOM BIRD

The same duffel bag I came to Vietnam with was packed and ready to go.

I included very few mementos of my time in the war. I packed the remnant partial casing of an exploded 122mm rocket that almost killed me in Quang Tri. The enemy casing had the letters "CCCP" in Russian stenciled on the side. Courtesy of the "USSR".

I was still in combat mode. I could not consider myself to be safe until I was on that Freedom Bird and out of South Vietnamese airspace. Anything could happen before then.

I caught a C-130 to Da Nang and then another hop to Saigon.

I checked into the transient barracks at Tan Son Nhut airbase. My Freedom Bird would depart the next day.

I ran into another guy heading home. We went to a bar just outside the base. He was kind of a strange dude. He said he was an MP from near the DMZ, close to Quang Tri. I noticed he was still packing a 45 caliber pistol under his civilian shirt.

What put me on alert was when he started talking in fluent Vietnamese with the luscious cocktail server.

The bar was full of armed military guys, so it appeared to be secure. As the booze flowed and talk got loose, I asked him how many tours he had served in Vietnam. He muttered, "a few".

I realized this guy was a lot more than a run-of-the-mill MP.

I thought he was C.I.A. or some kind of spook who had been all over Southeast Asia.

There was a covert side to the war. On a couple of occasions, I saw an all-white Huey slick refueling at the POL in Chu Lai. The slick had a name stenciled on the vertical fin, "Air America".

This was many years before I learned of the extent of the

C.I.A.'s involvement in the Southeast Asia war.

I tried to approach the "Air America" crew chief when he was refueling, but he avoided contact with me.

After a couple more drinks with my "MP" buddy, I called it a night. He was busy wooing the cocktail server in Vietnamese. It looked like he had won her over.

I made it back to base and hit the rack. The next morning, I got some breakfast, grabbed my duffel bag, and went over to out-processing.

The guy that was processing me asked if I had any weapons or ammunition. He added it was a criminal offense to have any of the contraband in my duffel bag.

I quickly confessed. I pulled out the three-foot-long rocket casing remnant and surrendered it. I would not risk screwing up my escape out of Vietnam.

My flight was finally called. As I stood in line for the check-in, my thoughts wandered over the last year. I still did not feel safe. I worried about a sapper barging into the terminal and detonating an explosive satchel. I had turned in my weapon, so I had no protection.

I made it through check-in and started walking out to the ramp towards the civilian airliner. I looked around nervously, searching for a sniper that could put a bullet in my head before I got into the plane.

I made it to the flight stairs and walked into the plane's cabin. I noticed fear and apprehension in the eyes of my fellow passengers.

There was no happiness. Everyone was just intent on getting airborne.

They finished loading the plane. The flight attendants closed the aircraft doors. There was little conversation.

The cockpit crew started up the engines and taxied to the active runway.

I thought about how a dink could fire an RPG at the plane as we crossed the airport perimeter at a low altitude.

The aircraft started its take-off run. We lifted off and gained altitude. I glanced out the window, looking for any rockets coming our way.

A few minutes later, the Captain keyed the intercom. "We have just cleared South Vietnam airspace".

Instantly, a unanimous cheer and jubilation arose from the passengers.

As the celebration died down, the soldiers realized they were safe.

The feeling was overwhelming. For the first time in many months, the constant cloud of lethality lifted. The emotions knew no bounds. We all thought about the soldiers who did not make it.

Soldiers sobbed and broke down. Flight attendants rushed to the more severely stricken soldiers to offer comfort and solace.

I sat in my seat, tensing tighter than ever before. I fought to rein in the overwhelming emotions that consumed my soul.

Everyone settled down.

There was nothing but silence on the plane as soldiers began to cope with life after the war.

It was early evening when we arrived at Fort Lewis, Washington.

The arrival was routine. Nothing special to welcome home soldiers who had put everything on the line for their country, day after day.

Everything was so strange. Yet another person inside me stirred, who I remembered, but lost touch with so long ago.

Before checking into the transient barracks for some sleep, I asked where I needed to go for that long-awaited "welcome home" steak dinner. They gave me directions to the mess hall.

As I walked up to the building, everything was dark. I found a sign on the door. "Mess Hall closed".

My flight to Wichita was scheduled for the next morning.

No "Welcome Home" steak dinner for me.

It may have never existed.

It began to rain.

As I walked to the barracks, I could feel some dejection.

I went to bed hungry.

My last thought before sleep was, "Thank God I am still alive."

December 26, 1971—*The 116th Assault Helicopter Company was withdrawn from Vietnam. The unit served in the Vietnam War from 1967 to 1971. The 116th was awarded the Presidential Unit Citation, four (4) Valorous Unit Citations, and the Meritorious Unit Commendation. All soldiers who served in the 116th were awarded the unit citation, Vietnamese Gallantry Cross.*

December 31, 1971 - *The number of U.S. military personnel in*

South Vietnam totaled 156,800. The highest American military strength was 549,500 in 1968.

August 11, 1972 - *The last combat unit in Vietnam, Third Battalion, Twenty-First Infantry, departed for the United States.*

August 30, 1975 - *The South Vietnamese stronghold of Saigon* (now known as Ho Chi Minh City) falls to the People's Army of Vietnam (NVA) and the Viet Cong. The South Vietnamese forces had collapsed under the rapid advancement of the North Vietnamese.

CHAPTER 53 SOLDIER'S AFTERMATH

The years after the war were difficult. The challenges for combat veterans to readjust to civilian life were known but not understood. Family and friends could not comprehend the torment we felt.

I never felt much remorse for all the people I killed. Most were trying to kill me and my crew, so I have been able to rationalize their deaths.

I wonder about the non-combatants that were collaterally killed or wounded by my bullets.

I am glad I did not kill that young boy who looked up at me with terror as my bullets hit around his feet. If I had killed that child, it would have been difficult to live with.

Sometimes, I wonder whether the boy survived the war and if he grew up to be an adult with a family. I hope he did.

America's policy on the War was not sound, and we did not make the correct strategic decisions.

Instead, our involvement degenerated into chaos. It deprived soldiers who served, the sense of accomplishment.

I know some veterans have a different view of the Vietnam War. They believe we fought and sacrificed for the cause of freedom.

I agree with that perspective to a point. Patriotism and passion to serve our country were my primary motivations for struggling to be assigned to combat in Vietnam.

History is the great equalizer. Facts rule the day.

As a consolation, I just remember the unspoken courage required for each of us to climb aboard those gunships every day.

It may have appeared to my family, the guy that went to war was the guy that came back.

At a deep, subconscious level, the war was still raging.

Despite my affliction of a tortured mind and an altered psyche, I honored my commitment to marry my girlfriend after my return from the war.

I had second thoughts just before the ceremony.

I knew I was not right. I was confused and conflicted.

Everyone was waiting inside the church as I languished in my car.

I proceeded with it and hoped I would recover.

I would never completely recover.

The marriage lasted for about ten years. We produced two daughters. I am sure I was responsible for the divorce.

The immediate post-war symptoms were bouts of anger that could be triggered by any kind of threat or confrontation.

I would retaliate against people who physically threatened me. I was conditioned in war to respond to threats with lethal force. I did not own a gun for several years after the war.

I have been an entrepreneur at heart since childhood. My military experience helped me to achieve business success. That drive to survive the Vietnam War transcended into the drive to make my business ventures successful. That entailed a lifetime of eighty to one hundred-hour work weeks. Still, the drive to succeed became my primary passion.

I did not regain a sense of humanity for decades. This made romantic relationships difficult, if not impossible.

I fulfilled my dream of becoming a pilot. I earned my Private Pilot's license on November 7, 1981. In the ensuing decades, I flew hundreds of hours all over America and the Caribbean.

I married for a second time. After the second marriage failed, I gave up on that institution.

The American Psychiatric Association recognized post-traumatic stress disorder in 1980. P.T.S.D. became widely known, primarily because of the difficulties experienced by Vietnam combat veterans.

I have been to the Vietnam Memorial in the Washington D.C. area twice. The first time, in the late 1980s, I could not endure the memories the Wall evoked. There was a group of vets dressed in jungle fatigues centered in an old Army tent on the Memorial grounds. They were welcoming Vietnam vets. One guy approached me. He could see I was suffering. He respected my privacy and allowed me to pass with my wife. Our eyes met briefly. I sensed he understood my pain as he nodded.

My wife and I left after a brief visit, even though it took many hours to travel to the Memorial.

The public disdain for the Vietnam experience persisted until after the Gulf Wars started in 1991.

I like to think America's leadership learned from the Vietnam experience and made military morale, specific military objectives, and public relations a top priority in the Gulf wars.

The American public responded and now embrace our military men and women. As the public renewed patriotic support of our military, Vietnam veterans are slowly being recognized.

Thank you.

In the 1990s, I was still having violent outbursts triggered by confrontations. I could not talk about the war without breaking down emotionally. I could not understand why I was still so severely affected by events that occurred over twenty years in the past.

I found a veterans outreach resource that was operated independently from the Veterans Administration or any government agency. My meeting with the counselor was explosive. The dam of pent-up emotions broke. The counselor asked me what I did during the war. I told him I was a helicopter gunship crew chief/door gunner. The counselor smiled and said, "No wonder you are having problems. Your job was one of the most dangerous in the Vietnam War." The counselor explained what was happening to me on a subconscious level.

I journeyed back to the Vietnam Memorial, alone, in the late 1990s. The experience was still painful.

I am relieved to say time can heal most wounds. Although the sound of my M-60 and rockets firing, incoming enemy rockets exploding, and images of the dead and dying are tucked away in a deep recess of my subconscious mind.

Maybe I will try to visit the Vietnam Memorial again, one more time.

In summation, I have no regrets.

I was proud to serve.

EPILOGUE

The Young answered the call of duty
for their divided country.
Their training was sound. The objective was not clear.
Their eyes were wide and full of excitement.
Most stepped forth with patriotism in their hearts.
They did not know the horrors that were to come.
They bid farewell to their loved
ones as caring Americans.
As the world turned against these duty-
bound warriors, they soldiered on.
Every major battle Americans fought was won.
The shadows of men they once were returned home.
They sought nurturing and warmth from
the bosom of their beloved country.
But they were turned away.
If our country survives for a thousand years,
may we not forget the service of Vietnam veterans.
Never again should we send our Young to war
in the absence of irrefutable cause and objective.

Specialist Fifth Class Fred Allen
United States Army 1969—1972

THANK YOU!

I hope you found Stingers to be a worthwhile reading experience. I would consider it a great honor if you would post your review of the book on the Amazon Stinger page.

I look forward to seeing your review!

Thank you for reading,

Stingers
Vietnam War - Helicopter Gunships

Fred Allen

USABooks1@aol.com

Made in the USA
Las Vegas, NV
16 December 2022

62913246R00128